# FUNDAMENTALS OF THERMOELECTRICITY

T0202083

# Fundamentals of Thermoelectricity

Kamran Behnia

*École Supérieure de Physique et de Chimie Industrielles, Paris, France*

# OXFORD

UNIVERSITY PRESS

Great Clarendon Street, Oxford, OX2 6DP,
United Kingdom

Oxford University Press is a department of the University of Oxford.
It furthers the University's objective of excellence in research, scholarship,
and education by publishing worldwide. Oxford is a registered trade mark of
Oxford University Press in the UK and in certain other countries

First published 2015
First published in paperback 2019
Impression: 1

Published in the United States of America by Oxford University Press
198 Madison Avenue, New York, NY 10016, United States of America

British Library Cataloguing in Publication Data
Data available

Library of Congress Cataloging in Publication Data
Data available

ISBN 978–0–19–969766–3 (Hbk.)
ISBN 978–0–19–884794–6 (Pbk.)

Printed and bound by
CPI Group (UK) Ltd, Croydon, CR0 4YY

*A Adèle, son électricité et sa chaleur*

# Preface

As a fundamental concept, thermoelectricity is both mature and under-explored. As a technology, it is disappointingly under-realized. Such a context leaves lots of room for research and for misunderstanding. Finding new materials with a large thermoelectric figure of merit is the subject matter of the investigation pursued along several orientations by a currently expanding community. In this quest, theoretical understanding may lag decades behind fortunate empirical discoveries. Future discoveries of new thermoelectric materials may depend more on serendipity than on advances in theoretical understanding. Such a possibility can be easily imagined by anybody familiar with the history of high-temperature superconductivity.

This application-oriented research is the focus of a number of books devoted to thermoelectric materials. *Introduction to Thermoelectricity* by Julian Goldsmid, a founding father of the field, is probably the best of them. Goldsmid discovered bismuth telluride, which has remained since 1953 the best-known thermoelectric material at room temperature.

The focus of the present book, on the other hand, is the concept and not its utility, at least not in the technological sense. D. K. C. MacDonald's little book of 1962, which devoted many pages to the puzzling thermoelectricity of elemental lithium, is the natural predecessor to this one. Written as an introduction to thermoelectricity for condensed-matter physicists, this book will try to address the following question: what kind of information on the electrons' organization in solids is yielded by measuring their thermoelectric response?

The first three quarters of the twentieth century witnessed the spectacular success of the band theory of solids in distinguishing between those solid bodies which conduct electricity and those which do not. During the last quarter, the attention gradually shifted to those cases where the band theory fails, presumably because of neglecting the Coulomb interaction. This shift of focus in attention has been accompanied by an increasing diversity in the experimentalist's toolbox. Good old bulk probes of matter are no more our only window to the behaviour of electrons in solids. In contrast to their predecessors, the current generation of solid-state physicists can probe the electron spectrum at the interatomic distances and measure the angular distribution of wave-vectors for electrons with different energies.

On the other hand, the central challenge to the condensed-matter physics remains a satisfactory description of the bulk properties of a given solid body. Thermoelectricity is one such property, traditionally classified as a transport property. Transport properties are notoriously more difficult to interpret than thermodynamic properties. After all, their analysis implies assumptions on the way travelling electrons are scattered. In thermoelectric transport, there is an entanglement between the propagation of electrons and the flow of energy in a solid body. No wonder then that over the years this less common transport property has acquired a double reputation for complexity as a subject of interpretation and for sensitivity as a probe of phase transitions.

However, this picture needs a serious correction. Consider the following question: what is the charge conductivity of a given solid body (say a piece of copper of a given shape and size) at zero temperature? The well-established answer to this question is rather disappointing: 'It depends on the quantity of impurities in the sample. They set the mean-free-path of electrons at low enough temperature.' In other words, there is no such thing as the bulk conductivity of copper at zero temperature. The cleaner the copper in question is, the lower its electric conductivity. On the other hand, if a similar question is asked about the thermoelectric conductivity, at least in principle, a quantitative answer is expected. In other words, in the zero-temperature limit, the magnitude of the Seebeck coefficient is not set by the presence of unavoidable extrinsic defects in the crystal. As we will see in Chapter 1, this fundamental feature is intrinsic to Callen's 1948 definition of the Seebeck coefficient. However, only recently have researchers on correlated electrons became aware of it.

In his paper, Callen distinguished between two components of the flow of heat in a solid body: one that is driven by a thermal gradient, and the other, by particle flow. The magnitude of the first is set by the thermal conductance of the solid in question, and the second, by its Seebeck coefficient. In the absence of a thermal gradient, the first component vanishes and only the second survives. In this situation, the Seebeck coefficient becomes simply the ratio of entropy flow to particle flow and thus a measure of entropy carried by travelling electrons. This fundamental conclusion based on thermodynamics has consequences which have not been appreciated as they would deserve. Consider heavy electrons, which acquire their mass by living in an environment rich in entropy. When they travel, the entropy they carry is not seen in a thermal conductivity measurement, but in their thermoelectric response. Most students of condensed-matter physics know that thermodynamic entropy is measured in a specific-heat experiment, but many fewer know that in its mobile version, this entropy shows up in the Seebeck coefficient and, in contrast, not in thermal conduction.

In the last three decades, new avenues have been opened in condensed-matter physics by considering conduction as transmission, an idea first put forward by Landauer. Take the question raised above about copper's electric conductivity. A contemporary answer to the question would invoke the quantum of conductance, $h/e^2$, and two material-dependent length scales: the Fermi wavelength of electrons in copper and their mean-free-path. What about the thermoelectric response? Unsurprisingly, most students and researchers in condensed-matter physics are unfamiliar with the answer. As discussed in Chapter 4, the relevant length scale for the thermoelectric response is the de Broglie thermal wavelength, which quantifies the thermal fuzziness of a Fermi sea and the physical uncertainty that the electrons can carry with them.

This book is intended to give an account of our current understanding of thermoelectric phenomena in solids by presenting basic theoretical concepts and numerous experimental results. It comes out that, even in the case of those simple metals considered to be domesticated long ago by quantum theory of solids, our understanding of thermo-electric response lags far behind known experimental facts. A recurrent theme of the book concerns forgotten puzzles.

To many readers of this book, it should be a surprise to learn that a consistent and unified theory for phonon drag is still missing. As we will see in Chapter 3, there are two historic theories of this phenomenon. In the picture devised for semiconductors by Herring, a phonon energy flow emerges as a response to electron collision. In an alternative picture developed for metals by Bailyn and others, an electronic flow is produced through interaction with phonons. It goes without saying that the two pictures should lead to an identical result. Amazingly, however, this is not the case, since there is no trace of a theoretical attempt to unify the two theories or to prove that one of them contains a fatal flaw as Sondheimer suspected decades ago.

Three chapters are devoted to a survey of experimental facts aim to revive a number of forgotten puzzles. The positive Seebeck coefficient of noble (and some alkali) metals has ceased to be 'a nagging embarrassment to the theory of the ordinary electronic transport properties of solids', in the words of Robinson written in 1967. But the embarrassment has vanished thanks to our forgetfulness and not to our cleverness. Both the puzzle and the solution proposed by Robinson have been largely forgotten. If his solution happens to be correct, however, it will have important implications for the interpretation of the sign of the Seebeck coefficient in many other systems.

Even more enigmatic than the positive Seebeck coefficient of noble metals at room temperature is their thermoelectric response at very low temperatures. Striking experiments at low temperature by Rumbo (see Fig. 6.7) have demonstrated that, at a temperature as low as 1 K, the sign and the magnitude of the Seebeck coefficient in copper is set by scattering details. On the other hand, and ironically one should say, it is in heavily doped semiconductors such as metallic silicon (see Fig. 6.23) and metallic $SrTiO_3$ (see Fig. 8.20) that experiment has quantitatively found what is theoretically expected in a simple Fermi liquid. Why is it so? Is it because in these latter cases, electron scattering has no strong structure in energy?

These are not the only forgotten puzzles. Before beginning to write this book, I did not know that there is a three-orders-of-magnitude gap between theory and experiment regarding the thermoelectric response of Bogoliubov quasi-particles of a superconductor. This is extremely disturbing, since theory seems to be a straightforward extension of the by Bardeen, Cooper, and Schrieffer theory and there is no such discrepancy in the case of thermal transport in the superconducting state. Moreover, the case of longitudinal thermoelectric response below critical temperature is to be contrasted to the case of transverse thermoelectric response above. The agreement between theory and experiment regarding the Nernst signal of fluctuating Cooper pairs in the normal state is excellent. Why? Is it just because it is more straightforward to carry out the experiment in the normal state?

Why such facts have gradually faded from the collective memory of condensed-matter physics is another question that deserves to be raised but is not addressed by this book. One of its aims, however, is to revive these puzzles and to stimulate fresh reflection and investigation on them. Another is to put side by side known facts, which are not often discussed alongside each other. The Seebeck coefficient in an underdoped cuprate and in a narrow-gap semiconductor when the carrier concentration in both is of the order of $10^{20}$ cm$^{-3}$ is similar enough in both magnitude and temperature dependence to temper any statement on the strangeness of one and the banality of the other.

The reader should not forget that this book is written by an experimentalist. It focuses more on metals (than on semiconductors), more on the low-temperature limit (than on room temperature and above), more on bulk materials (than on thin films) and more on curiosity-driven research (than on applications). These orientations reflect the author's background and expertise. They may also correspond to the cases where the confrontation between fundamental concepts and the available experimental data is more straightforward.

Paris, October 21, 2014

# Acknowledgements

I am particularly indebted to Jacques Flouquet, who introduced me to heavy electrons, and to Didier Jaccard, who taught me how to measure their thermoelectric response. Louis Taillefer has influenced my way of thinking about electrons in numerous ways during the past quarter of century. During this period, my research has benefited from interaction with numerous colleagues and students. Hervé Aubin, Aritra Banerjee, Romain Bel, Stéphane Belin, Lisa Buchauer, Cigdem Capan, Aurélie Collaudin, Benoît Fauqué, Xiao Lin, Saco Nakamae, Alexandre Pourret, Pana Spathis, Huan Yang, and Zengwei Zhu (who also extracted data for this book from numerous old published figures) are those who actively participated in the research leading to this book. I would also like to acknowledge stimulating discussions on thermoelectricity with Ana Akrap, Jean-Pascal Brison, Yuki Fuseya, Antoine Georges, Nigel Hussey, Yuji Matsuda, Kazumasa Miyake, Cyril Proust, and Phuan Ong. This book was partially written during my visiting professorships at the Tokyo Institute of Technology, the École Polytechnique Fédérale de Lausanne, and the University of Saõ Paulo, where I gave lectures on thermo-electricity to a public of graduate students; thus, I wish to thank my hosts in these three institutions: Koichi Izawa, Frédéric Mila, and Corlos Alberto dos Santos.

# Contents

# 1

# Basic Concepts

## 1.1 Electric and Thermal Conductivities

Some solid bodies permit electrons to pass through. These are conductors, which, as opposed to insulators, host mobile electrons. One can establish a charge current across such a body and measure the voltage generated by the applied current. The inverse procedure would be to impose an electric field along the conducting solid, and measure the flow of charge. Of these two equivalent procedures, the first is more practical. Conceptually, however, the second is easier to handle. Ohm's law is a statement on the link between the magnitude of the electric field and the charge flux density through a quantity dubbed conductivity:

$$\mathcal{J}^e = \sigma E \qquad (1.1)$$

The variable $\mathcal{J}^e$ is a measure of charge flow and represents the number of electrons travelling through a given cross section per second. The variable $E$, the electric field, is a measure of the spatial variation of the electric potential. Linking these two is the conductivity $\sigma$. A good conductor is a medium in which an ample charge flow accompanies a modest electric field. In an insulator, $\sigma$ is zero. One can establish a large electric field with no current passing through. In a superconductor, on the other hand, it is the resistivity, the inverse of conductivity, which vanishes. One can establish a finite $\mathcal{J}^e$ but no electric field.

Electric conductivity is a well-defined property of a solid body. It can be measured with great accuracy as a function of temperature and/or magnetic field. Such measurements have revealed much about the microscopic organization of electrons in the host solid. Superconductivity, for example, was discovered in 1911, when Kamerlingh Onnes tried to figure out what happens to the electric conductivity of mercury at very low temperatures.

Thermal conductivity is another property of a solid body. There is an obvious analogy with the case of electricity. Imposing a flow of heat in a solid generates a temperature gradient. Inversely, keeping a temperature difference along the body would create a heat current. The equivalent of Ohm's law for thermal transport is

*Fundamentals of Thermoelectricity*. First Edition. Kamran Behnia.
© Kamran Behnia 2015. Published in 2015 by Oxford University Press.

$$\mathcal{J}^Q = -\kappa \nabla T \tag{1.2}$$

The variable $\mathcal{J}^Q$ measures the amount of heat per second crossing a given section of the medium. $\nabla T$ is the temperature difference per unit length. Note the negative sign in this equation, which follows from the fact that heat flows from hot to cold, but the thermal gradient vector points from cold to hot. The symbol $\kappa$ is the thermal conductivity. Like $\sigma$, it is always positive. Unlike $\sigma$, on the other hand, it is always a finite number. Electric insulators lack mobile electrons, but their atoms can vibrate and therefore heat can flow through them. These elementary vibrations which can travel through the solid are called phonons. Because of their existence, we know of no solid body in which one can establish a zero $\nabla T$ and a finite $\mathcal{J}^Q$ or vice versa. Their capacity to carry heat over long distances is the reason why the best conductor of heat at room temperature is an electrical insulator, diamond.

Thermal conductivity is also a well-defined property of a solid body. Microscopically, both mobile electrons and phonons can carry heat. Therefore, it is harder both to measure and to interpret thermal conductivity as a function of physical parameters. Unsurprisingly, it has been studied less intensively than electric conductivity.

## 1.2   Entangling Heat and Charge

Thermoelectricity was discovered experimentally in 1821. That year, Thomas Johann Seebeck made a curious observation. After connecting a bismuth wire to an antimony wire, he found that by heating the junction between the two wires, he could create a voltage difference between the two free ends of the pair of wires. This was the first thermocouple. In 1834, Jean-Charles Peltier discovered the second thermoelectric effect. He found that one could generate heating or cooling by injecting at an electric current. Twenty years later, William Thomson (the future Lord Kelvin) argued that Seebeck and Peltier effects were intimately connected and, in fact, two manifestations of the same phenomenon.

Like electric and thermal conductivities, the thermoelectric conductivity is a well-defined property of a homogeneous solid body, in spite of the fact that, historically, it was discovered as a property of a junction between two materials. Usually, the experimental apparatus detects the difference in thermoelectric responses of two distinct interconnected materials. However, for the sake of clarity, let us conceptualize thermoelectricity in the context of a single homogeneous medium. The existence of such phenomena implies that the two above mentioned equations should be modified:

$$\mathcal{J}^e = \sigma E - \alpha \nabla T \tag{1.3}$$
$$\mathcal{J}^Q = \beta E - \kappa' \nabla T \tag{1.4}$$

The reason why $\kappa$ has become $\kappa'$ will become clear by the end of this chapter. The first of these two equations tells us that a charge current can be generated either by an electric field or by a temperature gradient. The second signifies that heat flows as a consequence of either a temperature gradient or an electric field. So, we have two new quantities, which we may call thermoelectric and electrothermal conductivities. According to Kelvin's argument, however, these two quantities are not independent:

$$\beta = \alpha T \tag{1.5}$$

This implies that there are only three (and not four) distinct conductivities: electric, thermal, and thermoelectric. The latter can be thought of either as the electric current produced by a temperature gradient, or the entropy current generated by an electric field. The two are identical. Lord Kelvin was not quite sure if Eq. 1.5 could be directly deduced from the laws of thermodynamics. He considered it a plausible conjecture and was happy to see that experimentation supported it. One had to wait till the fourth decade of the twentieth century to see the Kelvin relation based on a firm ground. In 1931, Lars Onsager used the principle of microscopic reversibility to deduce his reciprocal relations and showed that the Kelvin relation is one particular case of these relations. We will come back to the Onsager relations later.

The Seebeck coefficient $S$ is defined as the electric field generated by a thermal gradient in absence of a charge current:

$$S = E/(\nabla T) \tag{1.6}$$

Its units are volts per kelvin (V/K), but its magnitude rarely wanders above mV/K. The Peltier coefficient is defined as the heat flow produced by a charge flow:

$$\Pi = \mathcal{J}^Q/\mathcal{J}^e \tag{1.7}$$

Its units are volts. The original Kelvin relation states that

$$\Pi = ST \tag{1.8}$$

Both these quantities are easily accessible to the experimentalist, but this is not the case of the conceptually useful $\alpha$ (sometimes called Peltier conductivity). The latter can be determined by measuring both the Seebeck coefficient and the conductivity. The link between $S$ and $\alpha$ can be deduced by setting $\mathcal{J}^e = 0$ in Eq. 1.3:

$$S = \alpha/\sigma \tag{1.9}$$

The thermoelectric response can be positive or negative. This constitutes an important difference with thermal and electric conductivities, which are always positive. An electric field cannot generate a charge current of opposite direction: heat does not flow from cold to hot, but there is no fundamental constraint on the orientation of a charge current generated by a temperature gradient.

In anisotropic crystals, conductivity can be different along various directions. The values $\sigma$ and $\kappa$ can have different magnitudes along different orientations. In the case of $\alpha$, they can even be of different signs. In mathematical terms, the three conductivities are tensors and not scalars. This feature becomes flagrant in presence of a finite magnetic field.

## 1.3   Magnetic-Field-Induced Effects

Charged particles suffer a Lorenz force in the presence of a magnetic field. This force is perpendicular to their direction of propagation and to the magnetic field. Consider the flow of charge produced by an electric field. The existence of an additional Lorenz force causes the charge flow to deviate. As a consequence, $\mathbf{J^e}$ and $\mathbf{E}$ would cease to be parallel. In other words, one can measure a finite transverse electric field perpendicular to the orientation of the charge flow. This is the Hall effect, discovered in 1879 by the American physicist Edwin Hall. Eq. 1.1 can be reconciled with the existence of the Hall effect, provided that one replaces $\sigma$ with a matrix. In three dimensions, the two vectors $\mathbf{J^e}$ and $\mathbf{E}$ can have components along $x, y$, and $z$. The electric conductivity, which links them together through Eq. 1.1, is a $3 \times 3$ matrix:

$$\sigma = \begin{pmatrix} \sigma_{xx} & \sigma_{xy} & \sigma_{xz} \\ \sigma_{yx} & \sigma_{yy} & \sigma_{yz} \\ \sigma_{zx} & \sigma_{zy} & \sigma_{zz} \end{pmatrix}$$

In this matrix, components, which link the current and field along the same orientations (i.e. $\sigma_{ii}$) are called longitudinal conductivities, and those which relate components along two different orientations (i.e. $\sigma_{ij}$; with $i \neq j$) are called transverse or Hall conductivities. Only in the presence of a magnetic field may the off-diagonal (i.e. Hall) components of this matrix attain a finite value. The orientation of the magnetic field determines the non-zero off-diagonal component. If the magnetic field is along the $z$-axis, $\sigma_{xy}$ can become finite, as a current along $x$ can be associated with an electric field along $y$.

Since the Lorenz force would cause a flow of charged carriers carrying heat to deviate, it is not surprising that there is a thermal analogue of the Hall effect. The Righi-Leduc effect was independently discovered by the French physicist Anatole Leduc and the Italian physicist Augusto Righi in the 1880s. It is sometimes called the thermal Hall effect to stress the link with its better-known electric counterpart. In presence of a magnetic field, because of the existence of the Lorenz force, the heat-flow vector $\mathbf{J^Q}$ is not necessarily parallel to the temperature gradient $\nabla T$. To put this in mathematical terms, thermal conductivity $\kappa$ becomes a matrix with finite off-diagonal terms $\kappa_{ij}$ in the presence of a magnetic field.

The transverse thermoelectric effects are collectively known as the Nernst-Ettingshausen effect, as they were discovered by these two German scientists. Nowadays, one distinguishes between the Nernst effect (the off-diagonal counterpart of the Seebeck effect) and the Ettingshausen effect (which can be considered as the transverse Peltier effect). The Nernst effect is the generation of a transverse electric field (along $y$) by a longitudinal thermal gradient (along $x$) in the presence of a magnetic field (along $z$). The Nernst coefficient is thus defined as

$$N = E_y/(\nabla_x T) \tag{1.10}$$

The Ettingshausen effect, on the other hand, is the transverse thermal gradient generated by a longitudinal flow of charge in the presence of a finite magnetic field, and the Ettingshausen coefficient is defined as

$$\varepsilon = \nabla_y T/\mathcal{J}_x^e \tag{1.11}$$

The existence of transverse thermoelectric effects implies that the thermoelectric conductivity $\alpha$ is also a matrix like $\sigma$ and $\kappa$. Finite off-diagonal components such as $\alpha_{xy}$ emerge in the presence of a magnetic field. This off-diagonal Peltier component is a focus of much theoretical work on transverse thermoelectricity. However, it is only indirectly accessible to the experimentalist. It can be determined once both the Nernst and Seebeck coefficients as well as the longitudinal and Hall conductivities are known.

## 1.4 The Bridgman Relation

In 1924, Bridgman deduced a link between the Nernst and Ettingshausen coefficients through the thermal conductivity [Bridgman 1924]. According to the Bridgman relation:

$$N = \varepsilon\kappa/T \tag{1.12}$$

This relation is the counterpart of the Kelvin relation between the two longitudinal (Pelletier and Seebeck) thermoelectric coefficients. Bridgman's argument was purely based on thermodynamics. A succinct version of this argument was formulated by Sommerfeld and Frank in their 1931 paper [Sommerfeld and Frank 1931]. I do not see any reason to resist the temptation of reproducing this simple elegant argument.

Sommerfeld and Frank proposed the following thought experiment. Suppose that you measure the Ettingshausen coefficient of a sample in form of a rectangular cuboid (Fig.1.1a). The effect would lead to a temperature difference between the two lateral sides of the sample. One side becomes hotter than the other. Nothing would forbid you to try to use this temperature difference in a useful way. So, let us connect these two lateral sides to an external thermal machine which uses the thermal energy in one way or another. In this situation, heat would steadily flow along the $y$-axis and through the sample. Using Eq. 1.2, one can write

$$\mathcal{J}_y^Q = -\kappa\nabla_y T \tag{1.13}$$

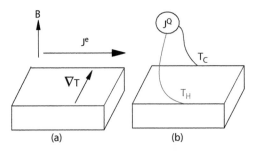

**Figure 1.1** *A thought experiment. (a) The Ettingshausen effect is the generation of a transverse thermal gradient by a longitudinal charge current in the presence of a magnetic field. (b) Sommerfeld and Frank proposed a thought experiment in which the temperature gradient is used by an external thermal machine.*

Note that we are assuming that thermal conductivity is isotropic. Now, the magnitude of the lateral thermal gradient is set by the amplitude of the primary charge flow and the size of the Ettingshausen coefficient:

$$\nabla_y T = \mathcal{J}_x^e \varepsilon \tag{1.14}$$

Therefore, the heat flow provided by the applied charge flow is

$$\mathcal{J}_y^Q = \kappa \mathcal{J}_x^e \varepsilon \tag{1.15}$$

This heat flow can be converted to energy in our external thermal machine. How much energy can we extract? The magnitude of energy per unit volume, $w$, produced by this heat flow is set by thermodynamics:

$$w = (\mathcal{J}_y^Q / T) \nabla_y T \tag{1.16}$$

Now, the conservation of energy implies that the cost of this energy should be entirely provided by the primary charge current injected into the sample. Let us have a closer look at the energy bill. Using the two previous equations, the energy per volume can be expressed as

$$w = (\mathcal{J}_x^e \kappa \varepsilon / T) \nabla_y T \tag{1.17}$$

This means that the unit volume of this solid body, with its specific $\kappa$ and $\varepsilon$ at a given temperature, is producing energy proportional to the injected charge flow. But this energy flow should be provided somehow. The Joule dissipation can take care of this. If we assume that establishing a lateral temperature gradient would always generate a finite longitudinal voltage, then the principle of the conservation of energy is saved. Now, this implies an assumption on the magnitude the Nernst coefficient, which links these two:

$$E_x = N\nabla_y T \tag{1.18}$$

Therefore, we can write

$$w = \mathcal{J}_x^e E_x = \mathcal{J}_x^e N\nabla_y T \tag{1.19}$$

Comparing Eq. (1.17) and (1.19) leads us to the Bridgman relation (Eq. 1.12).

In 1931, the same year that Sommerfeld and Frank published their version of the thermodynamic argument leading to the Bridgman relation, Onsager published his two celebrated articles on reciprocal relations [Onsager 1931a, b]. A few years later, Callen [1948] showed that the Bridgman relation can be algebraically deduced from Onsager reciprocal relations.

## 1.5   The Thermodynamic Origin of Thermoelectricity

At this stage, we can ask ourselves the following question: why thermoelectricity? What is the basic requirement for observing the generation of a thermal gradient by an electric field? Does it need a Fermi-Dirac distribution of the carriers? Do you need scattering? Would there be a Seebeck effect in a perfect crystal in which electrons can travel from one side of the crystal to the other?

The answers to these questions are contained in a paper which was published by Herbert Callen in 1948 and which gives a concise and elegant formulation of the Onsager reciprocal relations in the context of thermoelectricity [Callen 1948]. Moreover, Callen proposed a 'simple intuitive interpretation' (his words) leading to the formulation of an alternative definition of the Seebeck coefficient which is as rigorous as the one most commonly used. Here, we follow in his footsteps.

In equilibrium thermodynamics, an infinitesimal change in free energy, $\delta U$, can be expressed as

$$\delta U = T\delta\widetilde{S} + \mu\delta N + p\delta V \tag{1.20}$$

This equation contains extensive quantities, that is, entropy, $\widetilde{S}$, particle number, $N$, and volume, $V$, as well as intensive quantities, that is, temperature, $T$, chemical potential, $\mu$, and pressure, $p$. The extensive parameters scale with the size of the system, while the intensive ones do not. According to this equation, there are different ways to enhance the free energy in a given environment: increasing entropy, introducing new particles, or simply extending volume. In fact, you can define temperature, chemical potential, and pressure just as measures of the gain in energy by these alternative procedures. At this point, let us abandon the volume/pressure pair, as it does not play a role in the generation of thermoelectricity.

How can energy flow in such a system? How can it be carried from one place to the other? Let us quantify the energy flux $\mathcal{J}^U$ as the amount of energy flowing per second and unit area:

$$\mathcal{J}^U = T\mathcal{J}^S + \mu\mathcal{J}^N \tag{1.21}$$

$\mathcal{J}^{N,S}$ are also flux densities of particle number or entropy per second and per unit area. What this equation is telling us is that there are different ways of carrying energy from one place to the other. You can do it by displacing a number of particles, each with their chemical potential, or by moving a finite entropy (i.e. a finite number of configurations) at a given temperature. You can imagine that a black box containing either of them or a combination of both is flowing (See Fig. 1.2). The cost of the displaced energy can be calculated by translating particle number or entropy contained in the displaced black box into energy.

We are now interested in knowing how the system will respond to any spatial variation in temperature or in chemical potential. Let us take the divergence of the sides of the previous equation:

$$\nabla \cdot \mathcal{J}^U = (\nabla T) \cdot \mathcal{J}^S + T\nabla \cdot \mathcal{J}^S + \nabla\mu \cdot \mathcal{J}^N + \mu\nabla \cdot \mathcal{J}^N \tag{1.22}$$

Now, according to the first law of thermodynamics, energy is conserved. To fulfil this requirement, in any given region of the sample, the flow of incoming and outgoing energy should cancel out. This is a statement on the divergence of $\mathcal{J}^U$:

$$\nabla \cdot \mathcal{J}^U = 0 \tag{1.23}$$

If the number of particles is conserved, that is, if there is no local generation of particles, then

$$\nabla \cdot \mathcal{J}^N = 0 \tag{1.24}$$

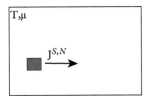

**Figure 1.2** *The concept of energy flow. At a given temperature and chemical potential, energy flows as a black box containing either entropy, particles, or a combination of the two.*

On the other hand, entropy is *not* conserved. In contrast to the free energy $U$ and the particle number $N$ in a steady-state system, entropy $\tilde{S}$ can grow with time. Let us call $\dot{s}$ the rate of change in entropy per volume; then,

$$\nabla \cdot \mathcal{J}^S = \dot{s} \tag{1.25}$$

With all these constraints, Eq. 1.22 can be rewritten as

$$0 = (\mathcal{J}^S \cdot \nabla T) + T\dot{s} + \mathcal{J}^N \cdot \nabla \mu \tag{1.26}$$

This equation deserves a closer look. It gives an account of how the flows of inward and outward energy end up by cancelling out. There are three components to the overall energy flow. The first term on the right represents the entropy flow riding on a temperature gradient, the second is the entropy locally produced, and the third is the particle flow times the decay of the chemical potential.

For the sake of simplicity, let us first focus on the case where no entropy is produced as time goes on. Then, the energy balance becomes simple. If we impose a gradient on the chemical potential, then particles moving along or opposite to this gradient would lose or gain potential energy as they flow. But as each particle has a finite degree of freedom, this flow of particles is also a flow of entropy. In order to conserve the total energy, a temperature gradient emerges. The sign and magnitude of this temperature gradient are set by the need to eliminate the total energy flow. In any system where both energy and particle number is conserved, a gradient in the chemical potential generates a temperature gradient (See Fig. 1.3). This is the fundamental reason behind the existence of thermoelectric phenomena.

Of course, the specific case of $\dot{s} = 0$ is very limited. It occurs when carriers travelling from one side of the solid body to the other do not suffer any change in their initial configuration. In almost all macroscopic solids, this would not happen; particles are scattered on their way and, therefore, generate local entropy. In this general case, the temperature gradient is set by the balance between the change in the potential energy and

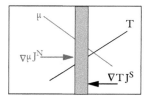

**Figure 1.3** *The origin of thermoelectricity. If both energy and particle number are conserved, then a particle flow riding on a chemical potential will unavoidably generate a temperature gradient.*

the steady production of entropy. The point is, however, that *the existence of thermoelectric phenomena is a consequence of the first law of thermodynamics in a system with a fixed number of particles* and does not result from local entropy production due to scattering events.

## 1.6   The Onsager Reciprocal Relations

'When two or more irreversible transport processes (heat conduction, electrical conduction and diffusion) take place simultaneously in a thermodynamic system, the processes may interfere with each other.' This is the starting sentence of Lars Onsager's first 1931 paper [Onsager 1931a], which set the foundations of a new science: the thermodynamics of irreversible processes. Onsager's approach not only put the Kelvin relation on a firm ground but, as seen above, led to the discovery of, fundamental reason behind the existence of thermoelectric phenomena.

The reciprocal relations revealed by Onsager are not limited to thermoelectrics. They are relevant to any case of two irreversible processes occurring simultaneously. In addition to the laws of thermodynamics, Onsager introduced an additional assumption, according to which, in the absence of a magnetic field, all microscopic processes are reversible: the future is symmetric to the past. This is called time-reversal symmetry.

To see a simple example of how microscopic time reversibility can generate a link between macroscopic transport coefficients, consider the case of Hall conductivity. As we saw in the previous section, off-diagonal components, $\sigma_{ij}$, emerge in the matrix $\sigma$ in the presence of a finite magnetic field. Time-reversal symmetry of microscopic processes implies that

$$\sigma_{ij}(B) = \sigma_{ji}(-B) \tag{1.27}$$

It is simple to see why. Let us assume that all carriers move backwards instead of forwards without altering the physics. If the charge flows backwards, that is, if the current vector points towards $-x$ instead of $x$, then inverting the orientation of the magnetic field towards $-z$ would suffice to create the same configuration of the three relevant vectors (current flow, electric field, and magnetic field). However, this latter configuration, in which current flows along $-x$ and the electric field along $+y$, is equivalent to the one in which the current is along $y$ and the electric field is along the $x$-axis. In other words, the configuration $(-x, y, -z)$ is equivalent to $(y, x, z)$; hence Eq. 1.27.

Onsager's insight was to generalize this feature to all 'kinetic coefficients'. In his 1948 paper, Callen showed that the Bridgman relation can be derived from reciprocal relations, as is also the case for the Kelvin relation.

In Callen's treatment, Eq. 1.22 is rewritten in term of the heat flux density, $\mathcal{J}^Q$, which is linked to the entropy flow as

$$\mathcal{J}^Q = T\mathcal{J}^S \tag{1.28}$$

It is easy to see that Eqs. 1.22–1.24 lead us to

$$\nabla \cdot \mathcal{J}^Q = -\nabla \left( \mu \cdot \mathcal{J}^N \right) \tag{1.29}$$

The left side of this equation represents the rate of heat production. The right side quantifies the loss of potential energy. The equation is telling us that the heat current emerges as a flow of kinetic energy to counter any change in the potential energy. As discussed above, this is a consequence of the principle of energy conservation.

Now, the rate of entropy production is set by

$$\dot{s} = \nabla \cdot \frac{\mathcal{J}^Q}{T} \tag{1.30}$$

Therefore,

$$\dot{s} = \mathcal{J}^Q \cdot \nabla \frac{1}{T} + \frac{1}{T} \nabla \cdot \mathcal{J}^Q \tag{1.31}$$

This equation tells us that there are two distinct sources of entropy production. The first term on the right side represents the entropy produced at the flow of heat from hot to cold. The second term represents the entropy produced at fixed temperature by the spatial variation of the heat flow. Using Eq. 1.29, this leads us to

$$\dot{s} = \mathcal{J}^Q \cdot \nabla \frac{1}{T} - \frac{1}{T} \nabla \mu \cdot \mathcal{J}^N \tag{1.32}$$

We have reached a point where the local production of entropy is expressed in terms of forces and fluxes, two fundamental concepts in non-equilibrium thermodynamics. In this picture, the rate of entropy production can be expressed as

$$\dot{s} = \sum_j \gamma_j \dot{\alpha}_j \tag{1.33}$$

$\gamma_j$ is a force and $\dot{\alpha}_j$ is its associated flux (or flow). In our context, one can see by comparing the two equations that the two vectors $\mathcal{J}^Q$ and $\mathcal{J}^N$ are fluxes, and the forces are $\nabla \frac{1}{T}$ and $\frac{1}{T} \nabla \mu$. Forces generate fluxes. Or maybe fluxes imply forces? As discussed in the beginning of this chapter, we don't need to know which comes first. A diffusive movement implies a finite force and a finite flux. Kinetic coefficients link forces to fluxes. Now, taking $\mathcal{J}^Q$ and $\mathcal{J}^N$ as fluxes, and $\nabla \frac{1}{T}$ and $\frac{1}{T} \nabla \mu$ as forces, the kinetic coefficients can be defined as

$$\mathcal{J}^N = L_{11} \frac{1}{T} \nabla \mu + L_{12} \nabla \frac{1}{T} \tag{1.34}$$

$$\mathcal{J}^Q = L_{21} \frac{1}{T} \nabla \mu + L_{22} \nabla \frac{1}{T} \tag{1.35}$$

where, according to the Onsager relation,

$$L_{21} = L_{12} \tag{1.36}$$

It may not have escaped the reader's attention that the pair of equations 1.34–35 is reminiscent of 1.3–4. Whenever there is a constant amount of charge per particle, $e$, the particle flow is just charge flow times $e$. Therefore, $\mathcal{J}^e = e\mathcal{J}^N$. The kinetic coefficients $L_{11}$, $L_{12}$, and $L_{22}$ are obviously linked to the three conductivities $\sigma$, $\alpha$, and $\kappa$. The reciprocal relation expressed by Eq. 1.35 is the origin of the Kelvin relation.

Electric conductivity, $\sigma$, is the charge current flow produced by a gradient in the electric potential($\frac{1}{e}\nabla\mu$ ) in the absence of a thermal gradient. Therefore,

$$\sigma = \frac{e^2}{T}L_{11} \tag{1.37}$$

As for $\alpha$, it is easy to see that

$$\alpha = \frac{e}{T^2}L_{12} \tag{1.38}$$

The case of thermal conductivity is more complicated. It is important to distinguish between the thermal conductivity $\kappa$ in absence of the charge current ($\mathcal{J}^e = 0$) and the conductivity $\kappa'$, which occurs in the absence of a potential gradient ($\nabla\mu = 0$ ). The latter is obviously

$$\kappa' = \frac{1}{T^2}L_{22} \tag{1.39}$$

The former can be calculated from Eq. 1.33 by setting $\mathcal{J}^N = 0$:

$$\kappa = \frac{1}{T^2}\frac{L_{11}L_{22} - L_{12}^2}{L_{11}} \tag{1.40}$$

The reader is invited to check that

$$\kappa' = \kappa\left(1 + \frac{\alpha^2 T}{\sigma\kappa}\right) \tag{1.41}$$

The dimensionless ratio which appears on the right side of this equation, $(\frac{\alpha^2 T}{\sigma\kappa})$, is the thermoelectric figure of merit. This quantity has a deep significance. It sets a limit to any potential manipulation of thermoelectricity as an energy switch. We will come back to it later.

## 1.7  The Seebeck Coefficient as a Measure of Entropy per Carrier

The Seebeck coefficient of a given material is defined as the electric field generated by a temperature gradient. In contrast to those for electric or thermal conductivities, its units are the same in one, two, or three dimensions: volts per kelvin. Remarkably, it does not contain any reference to spatial dimensions. Why? Let us review the way it is actually measured.

The standard procedure for measuring the Seebeck coefficient is to impose a temperature difference along the sample and ensure the absence of any charge current. Then, by measuring both $\delta V$, the voltage difference, and $\delta T = (T_1 - T_2)$, the temperature difference, on two separate electrodes along the sample, one can determine the Seebeck coefficient of the sample at an average temperature of $T_{av} = \frac{T_1 + T_2}{2}$:

$$S(T_{av}) = \frac{\delta V}{\delta T} \tag{1.42}$$

Since the temperature gradient and the electric field share an identical length scale, the latter vanishes in the final result. The Seebeck coefficient does not depend on the distance between the electrodes or on the thickness or any other dimensions of the sample. The Seebeck coefficient, in contrast not only to charge conductivity and to other bulk properties of the system, does not scale with size. *It is an intensive quantity.*

In the pair of kinetic equations (Eqs. 1.3 and 1.4) describing the thermoelectric phenomena in a solid, the Seebeck coefficient does not explicitly appear. In this formal presentation, the explicit thermoelectric effect is represented by $\alpha$, the so-called Peltier conductivity. Measuring $S$ consists in solving Eq. 1.3 for $\mathcal{J}^e = 0$:

$$\frac{\alpha}{\sigma} = \frac{E}{\nabla T} \tag{1.43}$$

In other words,

$$S = \frac{\alpha}{\sigma} \tag{1.44}$$

Units of both conductivities contain a reference to sample dimensions: $\sigma$ is expressed in $\Omega^{-1}\,m^{-1}$, and $\alpha$ in $A\,K^{-1}\,m^{-1}$. However, these vanish in their ratio, the Seebeck coefficient, which does not contain any reference to length in its units. In spite of the impression given by our set of kinetic equations, $S$ can often provide a more intuitive access to the heart of the thermoelectric phenomena. As an example, we will see in a following chapter (in a Fermi liquid), the Seebeck coefficient is the most straightforward way to access the Fermi temperature of the system.

There is an alternative definition of the Seebeck coefficient that was formulated by Callen in his 1948 paper. This definition is rigorously equivalent to the more common one but is more instructive for many purposes. According to this definition, *the Seebeck coefficient is the ratio of entropy flow to particle flow in the absence of a thermal gradient.*

This can be seen by expressing $\mathcal{J}^Q$ as a function of $\mathcal{J}^e$ and $\nabla T$. Combining 1.34 and 1.35, one obtains

$$\mathcal{J}^Q = -\frac{L_{12}}{L_{11}}\mathcal{J}^N + \left(L_{22} - \frac{L_{12}^2}{L_{11}}\right)\nabla\frac{1}{T} \tag{1.45}$$

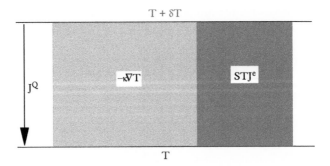

**Figure 1.4** *The two components of heat flow in the presence of charge flow. When a temperature gradient is established by a charge current, there are two components to the heat flow. The first is proportional to the magnitude of the Seebeck coefficient. The second is set by the amplitude of the thermal conductivity. The first component is the one associated with particle flow.*

In other words, there are two components to the heat current: one which is carried by the flow of particles, and another carried by the thermal gradient (see Fig. 1.4). Now, if we replace $L_{ij}$ with the more familiar kinetic coefficients $\alpha$, $\sigma$, and $\kappa$, we find that

$$\mathcal{J}^Q = \frac{\alpha}{\sigma} T \mathcal{J}^e - \kappa \nabla T \tag{1.46}$$

Or equivalently,

$$\mathcal{J}^S = S \mathcal{J}^e - \kappa \frac{\nabla T}{T} \tag{1.47}$$

This simple equation derived from first principles by Callen has deep implications. It tells us that one can set an entropy flow in the solid under question either by imposing a thermal gradient or by generating a current of electrons (or, of course, by a combination of both). The Seebeck coefficient quantifies the entropy flow proportional to the charge current and caused by the latter. The Seebeck coefficient is just a measure of the number of degrees of freedom a travelling particle carries with itself.

## 1.8  Figure of Merit

Figure of merit is a dimensionless ratio involving the three conductivities:

$$ZT = \frac{S^2 T \sigma}{\kappa} \tag{1.48}$$

It quantifies the thermoelectric performance of a given material. In a thermoelectric device, as far as the choice of the material plays a role, the figure of merit is what eventually sets the efficiency.

We saw above that one should distinguish between the two thermal conductivities $\kappa$ and $\kappa'$. The former is the thermal conductivity, the ratio of the heat current density to temperature gradient in the absence of charge current. The latter is the 'thermal conductivity', defined as the ratio of the component of heat flow *not* caused by electric field to the thermal gradient and which appears in Eq. 1.3. The two equations are related through the figure of merit:

$$\kappa' = \kappa(1 + ZT) \tag{1.49}$$

Consider the flow of heat across a material with no flow of charge. In this case, since $\mathcal{J}^e = 0$, the flow of heat is just $-\kappa \nabla T$. Now, because of thermoelectricity, there will be a finite electric field across the sample, and the heat flow will have two components: $-\kappa' \nabla T$ and $\alpha TE$. (see Fig. 1.5). $ZT$ sets the relative proportion of these two components. One of the two components constitutes a fraction of total heat flow proportional to E.

Since $\mathcal{J}^e = 0$, Eq. 1.3 implies that this electric field is equal to

$$E = \frac{\alpha \nabla T}{\sigma} \tag{1.50}$$

Hence, as one would expect, both components of the heat flow are proportional to the thermal gradient. The first one is $-\kappa' \nabla T$ and the other is $\frac{\alpha^2}{\sigma} T \nabla T$. Now, the ratio of the second component to the total heat flow $(-\kappa \nabla T)$ is set by

$$\frac{\alpha^2 T}{\sigma}/(-\kappa) = -ZT \tag{1.51}$$

Thus, the figure of merit quantifies the relative strength of the component of the heat flow set by the electric field. Note that this component flows opposite to the one going

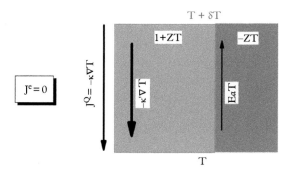

**Figure 1.5** *The two components of the heat flow in absence of charge flow. The heat flow has a component proportional to the thermal gradient and another one proportional to the electric field. The relative weight of these two components is set by the figure of merit $ZT$.*

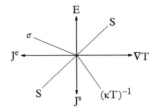

**Figure 1.6** *The four vectors.*
*Each vector is linked to its*
*immediate neighbour by a*
*conductivity scalar.*

from hot to cold. An important threshold is reached when $ZT$ becomes equal to unity. In this case, heat will still flow from hot to cold, but there is perfect compensation between what is mediated by the electric field and what is set by thermal conductivity.

The vectors $\mathbf{J}^Q$, $\mathbf{J}^e$, $\nabla\mathbf{T}$, and $\mathbf{E}$ are linked together by three conductivities. It may be instructive to display them as a set of four vectors, each normal to its immediate neighbour (See Fig. 1.6). In this representation, each of the four vectors is related to its neighbour to its right by a conductivity scalar. Note that the Seebeck coefficient, $S$, which links the entropy flux and charge flux, is the same which relates the electric field and the temperature gradient, as Onsager reciprocity would imply.

This geometric representation provides an interesting window to the meaning of the figure of merit, by drawing an analogy with the Hall effect.

In absence of a magnetic field, $\mathcal{J}_x$, the charge current density along the $x$-axis, is set by the electric field along the $x$-axis times the diagonal element of the electric conductivity, (i.e. $\sigma_{xx}E_x$). The same is true for the $y$-axis. In presence of a magnetic field and a finite Hall coefficient, this is no longer the case. The charge current vector is partly set by the product $\sigma_{xy}E_y$. In other words, it rotates (See Fig. 1.7b). This is also true for $\mathcal{J}_y$, which rotates by the same angle and remains perpendicular to $\mathcal{J}_x$. The angle, called the Hall angle, is related to the ratio of non-diagonal to diagonal components of the conductivity tensor.

Now, let us look at the thermoelectric phenomenon in a similar frame. If there were no thermoelectricity, the heat current $\mathcal{J}^Q$ would have been set by the thermal gradient and thermal conductivity ($\kappa\nabla T$). Similarly, charge current would have been solely determined by electric field and electric conductivity ($\sigma E$) (Fig. 1.7c). When there *is* thermoelectricity, on the other hand, the thermal current is partly set by the electric field and electric conductivity. This means that, in the ($\kappa\nabla T$, $\sigma E$) plane, $\mathcal{J}^Q$ has rotated. The charge current vector, now partly set by the electric field, suffers an identical rotation (Fig. 1.7d). In this picture, the figure of merit is what quantifies the magnitude of rotation. A $ZT$ of unity corresponds to a $\pi/4$ rotation of the currents with respect to their initial orientations. In this situation, both charge and heat current are set in half by the electric field and in half by the thermal gradient.

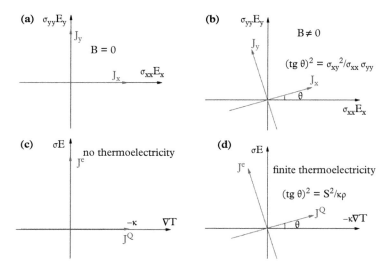

**Figure 1.7** *Figure of merit and the Hall angle. (a) In the absence of a magnetic field, current density along a given orientation is set by the product of the electric field and the diagonal component of electric conductivity along that orientation. (b) In the presence of a magnetic field, this is no longer the case, and the current density vector rotates with respect to its initial orientation. (c) In the absence of thermoelectricity, the heat current is set by thermal gradient and thermal conductivity and the charge current by the electric field and charge conductivity. (d) When there are thermoelectric phenomena, the two vectors rotate in a similar manner. The figure of merit quantifies the amplitude of this rotation.*

## 1.9 The Electricity in Thermoelectricity

As we saw in this chapter, thermoelectricity emerges as a consequence of energy conservation combined with particle conservation. It happens that the phenomenon was historically discovered in solids, where the particles in question are electrons, each with a determined electric charge. At this level of abstraction, the electricity in thermoelectricity is anecdotic. There is an irony here, since in a contemporary condensed-matter laboratory, the most crucial instrument for measuring the Seebeck coefficient with a high resolution is a good nanovoltmeter, with precautions taken to protect the signal from electric noise. But this does not have any profound conceptual consequences. An emerging field of investigation is the equivalent of thermoelectricity in cold atoms [Brantut 2013], where the particles in question are neutral and there is no flow of electric charge.

In the next chapter, we will address the specific case of thermoelectricity of electrons in a solid. As we will see, even then, what counts more than the electric charge of electrons is their fermionic nature and its implications for their statistical distribution.

# 2

# The Semiclassical Picture

In the previous chapter, we saw that thermoelectricity arises because of basic thermo-dynamic considerations. The fact that the particles in question are electrons flowing in a solid was entirely overlooked. In this chapter, we are going to focus on this.

## 2.1 The Fermi-Dirac Distribution

In statistical physics, particles of a gas have different energies according to a temperature-dependent distribution probability. In a classical gas, this is the Maxwell-Boltzmann distribution. The basic scheme is simple. It is less probable to find high-energy particles than to find low-energy ones. Moreover, warming up the gas in-creases the chance of finding high-energy particles. Quantitatively, at temperature $T$, the probability of finding a particle with energy $\epsilon$ is

$$f^{MB}(\epsilon) = e^{-(\epsilon-\mu)/k_B T} \tag{2.1}$$

The origin of the energy distribution is set at $\mu$, which is the chemical potential (i.e. the cost of adding a new particle to the system). According to this equation, particles with an energy larger than $\mu$ are exponentially less probable to be found. As the temperature decreases, it becomes more unlikely to find high-energy particles. At absolute zero, all particles have the same energy $\mu$.

It happens that, unlike particles of a classical gas, the electrons of a crystal cannot have the same energy. They are indistinguishable particles and their collective wave-function is expected to change sign whenever two particles exchange their places with each other. This is not possible if two identical electrons occupy the same quantum state. As a result, no two identical electrons possess the same energy and thus, the Maxwell-Boltzmann distribution does not apply to them. Instead, their energy distribution follows the Fermi-Dirac distribution. Accordingly, the probability of an electron with energy of $\epsilon$ is

$$f^0(\epsilon) = \frac{1}{e^{[(\epsilon-\mu)/k_B T]} + 1} \tag{2.2}$$

*Fundamentals of Thermoelectricity*. First Edition. Kamran Behnia.
© Kamran Behnia 2015. Published in 2015 by Oxford University Press.

At absolute zero, the Fermi-Dirac distribution becomes a step function. It is one for $\epsilon < 1$ and zero for $\epsilon < 1$. In other words, electrons would occupy all available states up to the chemical potential and leave the other states empty. This zero-temperature chemical potential is the Fermi energy $\epsilon_F$. At the extreme limit of high temperatures (that is, when the first term in the denominator is much larger than unity), it will become equivalent to the classical distribution of Eq. 2.1. In this case, thermal energy gives the opportunity for electrons to occupy distinct states and 'forget' their quantum nature. Between these two limits, the distribution presents a passage from occupied to unoccupied states, which becomes broader as the temperature increases (see Fig. 2.1).

The Fermi energy of an electron system, $\epsilon_F$, depends on the properties of the specific solid in which the electrons exist. Let us call the total number of electrons $N_{tot}$. An electron census can be done by taking into account the probability for an electron to have a given energy and the number of electrons allowed to have that energy. This yields

$$N_{tot} = \int \int_0^\infty f^0(\epsilon) N(\epsilon) d\epsilon \, dV \tag{2.3}$$

The electron density of states, $N(\epsilon)$, quantifies the number of states per unit of energy and per unit of volume. In the case of the free electrons, which are plane waves with a definite wave-vector, the density of states is proportional to $\epsilon^{1/2}$. The first integral is over

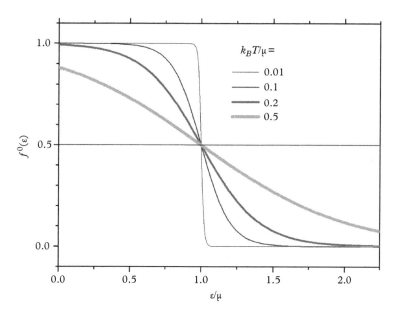

**Figure 2.1** *Thermal broadening of the Fermi-Dirac distribution. The Fermi-Dirac distribution (Eq. 2.2) for different temperatures. It becomes broader as the temperature increases. At different temperatures, there is a one-half probability of finding an electron with an energy equal to the chemical potential. However, the chemical potential moves to lower energies as $k_B T$ increases.*

the volume of the crystal. In a homogenous material, the carrier density is just $n = N_{tot}/V$ and the equation becomes

$$n = \int_0^\infty f^0(\epsilon) N(\epsilon) d\epsilon = \int_0^{\epsilon_F} N(\epsilon) d\epsilon \tag{2.4}$$

Let us consider first the case of a metal at absolute zero. There is a finite carrier concentration $n$ which sets the Fermi energy $\epsilon_F$. The cost of adding one electron to the system is the chemical potential, or the Fermi energy. At absolute zero, the two quantities are identical. However, as the system warms up and since the Fermi-Dirac function $f^0$ evolves with temperature, this would no longer be the case. One can use a procedure dubbed the Sommerfeld expansion [Ziman 1964] to estimate the temperature dependence of the chemical potential. It turns out that

$$\mu \approx \epsilon_F - \frac{\pi^2}{6}(k_B T)^2 \frac{1}{N(\epsilon_F)} \frac{\partial N(\epsilon)}{\partial \epsilon}\Big|_{\epsilon=\epsilon_F} \tag{2.5}$$

Thus, at a finite temperature, the chemical potential is lower than the Fermi energy. At a given temperature, the chemical potential is the energy at which the probability of finding an electron is one half. This energy scale is no more what it was at infinitesimal temperatures and shifts downwards with increasing temperature.

As a consequence, in the presence of a thermal gradient along the solid body, the chemical potential is lower at the hot side. It costs more energy to add an electron to the cold side than to the hot side. Naturally, the electrons are forced to migrate towards the hot side.

This opens an alternative window to the origin of thermoelectricity. With no net flow of electrons, a thermal gradient generates a voltage. This is because electrons would be subject to two compensating forces: a thermal force pushing it from cold to hot, and an electric force pushing it in the opposite direction.

Thus, thermoelectricity can be considered a consequence of the temperature dependence of the chemical potential, itself a result of the thermal broadening of the Fermi-Dirac distribution.

## 2.2  Two Formalisms: Boltzmann and Landauer

There are two distinct mathematical descriptions of the way electrons carry heat and charge through a solid body. The older and better-known approach is based on the Boltzmann equation. Historically, this has been the theoretical path to calculate the transport coefficients of bulk materials in the Drude picture [Ashcroft and Mermin 1976; Ziman 1960; Ziman 1964].

A second approach is called Landauer or Landauer-Büttiker formalism [van Houten *et al.* 1992]. Originally developed as an analysis of charge conductivity in one-dimensional conductors [Landauer 1957] and mesoscopic devices [Büttiker 1986], this approach was extended to treat thermoelectric transport [Sivan and Imry 1986;

Streda 1989; Butcher 1990; van Houten *et al.* 1992]. More recently, it has been employed to treat thermoelectric response in bulk materials as well [Jeong *et al.* 2010].

## 2.2.1 The Boltzmann Equation

The central concept in the Boltzmann approach is the existence of the function $f_k(\mathbf{r})$, which represents the probability of finding a carrier with the wave-vector $\mathbf{k}$ at the position $\mathbf{r}$. An electric field or a temperature gradient would affect this function by making the presence of electrons in particular places more probable. As the carriers travel across the sample or scatter with each other, they create variations in the magnitude of $f_k(\mathbf{r})$ at a given position, a given moment, and for a given wave-vector.

Each electron has a charge of $e$ and a group velocity of $v_k$:

$$v_k = \frac{1}{\hbar} \frac{\partial \epsilon_k}{\partial \mathbf{k}} \tag{2.6}$$

The travelling electrons would generate a flow of electric charge equal to $e v_k f_k$. The total charge flow is obtained by integrating over all wave-vectors:

$$\mathbf{J}^e = \int v_k e f_k dk \tag{2.7}$$

There is a corresponding expression for the energy flux. Since the thermal energy of each travelling electron is $\epsilon - \mu$, the heat flux is represented by the following integral:

$$\mathbf{J}^U = \int v_k \epsilon f_k dk \tag{2.8}$$

Using equation 1.21, one can extract the expression for the heat flux:

$$\mathbf{J}^Q = \mathbf{J}^U - \mu \mathbf{J}^N = \int v_k (\epsilon - \mu) f_k dk \tag{2.9}$$

In the absence of an electric field or a temperature gradient, the function $f_k(\mathbf{r})$ is identical to the Fermi-Dirac distribution $f^0$ of Eq. 2.2. Basically, the effect of an external field would be to skew the wave-vector distribution of electrons. If electrons begin to travel from one region to the other, they will generate a spatial variation in the magnitude of $f_k(\mathbf{r})$. Finally, there are scattering events which cause a change in the state of the electrons. The Boltzmann equation states that the overall result of all these three effects cancels each other out.

In order to work out a manageable equation, several approximations are in order. First of all, one usually assumes that the deviation from the Fermi-Dirac distribution remains in the linear regime. There is also the relaxation-time approximation, according to which, there is a typical time lapse $\tau$ between two scattering events:

$$\frac{df_k(\mathbf{r})}{dt}\bigg|_{\text{scattering}} = \frac{f_k(\mathbf{r}) - f^0}{\tau} \tag{2.10}$$

The variation in $f_k(\mathbf{r})$ caused by a temperature gradient can be expressed as

$$\frac{df_k(\mathbf{r})}{dt}\Big|_{Temp.grad.} = v_k \cdot \frac{\partial f_k(\mathbf{r})}{\partial \mathbf{r}} \approx v_k \cdot \frac{\partial f^0}{\partial T}\nabla T \tag{2.11}$$

The approximation consists in supposing that the effect of the variation in temperature is mainly supported by the thermal broadening of the Fermi-Dirac distribution. Between two scattering events, the electrons drift under the influence of the thermal gradient. A similar expression can be worked out for the electric field:

$$\frac{df_k(\mathbf{r})}{dt}\Big|_{Elec.field} = \frac{e}{\hbar}E\frac{\partial f_k(\mathbf{r})}{\partial \mathbf{k}} \approx ev_k \cdot \frac{\partial f^0}{\partial \epsilon_k}.E \tag{2.12}$$

The linearized Boltzmann equation would then become

$$v_k \cdot \frac{\partial f^0}{\partial T}\nabla T + ev_k \cdot \frac{\partial f^0}{\partial \epsilon_k} \cdot E + \frac{f_k(\mathbf{r}) - f^0}{\tau} = 0 \tag{2.13}$$

This means that the local distribution deviates from its steady state by two perturbations:

$$f_k(\mathbf{r}) = f^0 - \tau \cdot v_k \left(\frac{\partial f^0}{\partial T}\nabla T + e\frac{\partial f^0}{\partial \epsilon_k} \cdot E\right) \tag{2.14}$$

Let us pause and examine this equation. Two terms in the right side of this equation express the drift of electrons with electric field and temperature gradient. Both invoke the way Fermi-Dirac distribution responds to a perturbation, with a significant difference. In the case of the electric field, it is the energy derivative of the distribution and, in the case of the temperature gradient, it is the temperature derivative. Now, as seen in Fig. 2.2, a slight change in energy would rigidly shift the distribution, but warming up would broaden it. Thus, imposing an electric field would create a situation where electrons travelling along one direction become more energetic than those travelling in the opposite direction. In presence of a thermal gradient, on the other hand, the distribution is sharper at the cold side, and a thermal force emerges because of this difference in distribution width at the two extremes of the sample. Significant consequences for the physical properties of the system arise.

Let us calculate what the electric conductivity of the system would be in the absence of a thermal gradient. We should now combine Eq. 2.7 with Eq. 2.14 and set $\nabla T = 0$:

$$J^e = \int \left(v_k e f^0 - e^2 \tau v_k \cdot v_k \frac{\partial f^0}{\partial \epsilon_k} \cdot E\right) dk \tag{2.15}$$

Now, in the steady-state situation, the average velocity is zero. Therefore, the first term vanishes and using Eq. 1.3 (or 1.1) we obtain

$$\sigma = -e^2 \int \tau(k) v_k \cdot v_k \frac{\partial f^0}{\partial \epsilon_k} dk \tag{2.16}$$

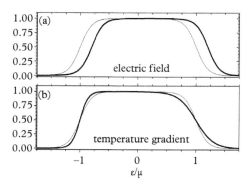

**Figure 2.2** *The Fermi-Dirac distribution in the presence of an electric field and a temperature gradient. (a) In presence of a finite electric field, the distribution rigidly shifts. (b) When a temperature gradient is imposed, the width of the Fermi-Dirac distribution is different for electrons travelling along and opposite to the temperature gradient.*

If we now set the electric field equal to zero, we can calculate the charge current generated by a thermal gradient, again using Eqs 2.7 and 2.14:

$$\mathbf{J^e} = -\int e\tau v_k \cdot v_k \frac{\partial f^0}{\partial T} \nabla T dk \tag{2.17}$$

This should be compared with Eq. 1.3, and it allows us to determine the magnitude of $\alpha$:

$$\alpha = -e \int \tau(k) v_k \cdot v_k \frac{\partial f^0}{\partial T} dk \tag{2.18}$$

The $k$-dependence of the scattering time is explicitly stated in Eqs 2.16 and 2.18. These two equations tell us how electric and thermoelectric conductivities are linked to each other, to the Fermi-Dirac distribution, and to electron velocity and relaxation time. The two latter quantities are specific to a given material and can be strongly anisotropic. Let us note two features about them.

First, there is the issue of sign. Unsurprisingly, charge conductivity $\sigma$ is always positive. The sign of all terms in the integral of Eq. 2.16 are known in advance: $\frac{\partial f^0}{\partial \epsilon_k}$ is always negative, and $\tau(k) v_k^2$ is always positive. On the other hand, $\alpha$ can be either positive or negative.

The second observation is that both quantities probe the average value of a combination of relaxation time and electron velocity over the entire $k$-space, with a significant difference. They don't use the same pondering factor for averaging. This pondering factor is the energy derivative of the distribution function in the case of charge conductivity, but the temperature derivative in the case of thermoelectric conductivity.

This means that the carriers which dominantly contribute to the two transport processes are not the same, as seen in Fig. 2.3. In the case of charge conductivity, those electrons which are precisely at the chemical potential are those which contribute most to the $k$-space average probed by $\sigma$. In the case of thermoelectric transport, these carriers are those who do not contribute at all. The overall signal is set by the contribution of electrons with an energy barely higher or barely lower than the chemical potential. Moreover, the two contributions have opposite signs.

In order to calculate the thermal current, one can start with Eq. 2.14 and, using the properties of the Fermi-Dirac distribution (Eq. 2.2), rewrite it in the following way [Ziman 1960, 1964]:

$$f_{\mathbf{k}}(\mathbf{r}) - f^0 = -\tau v_k \left[ -\frac{(\epsilon - \mu)}{T} \frac{\partial f^0}{\partial \epsilon_{\mathbf{k}}} \nabla T + e \frac{\partial f^0}{\partial \epsilon_{\mathbf{k}}} E \right] \tag{2.19}$$

Using this equation, we can express the heat flux using Eq. 2.9:

$$\mathbf{J}^Q = \int v_k (\epsilon - \mu) \tau v_k \left[ \frac{(\epsilon - \mu)}{T} \frac{\partial f^0}{\partial \epsilon_{\mathbf{k}}} \nabla T - e \frac{\partial f^0}{\partial \epsilon_{\mathbf{k}}} E \right] dk \tag{2.20}$$

Note that, like the electric conductivity, only the deviation from the steady-state distribution counts as the integral over the steady-state function. Using Eq. 1.4, one

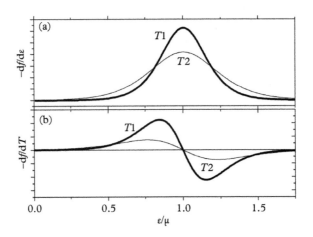

**Figure 2.3** *Carriers contributing to electric and thermoelectric transport. (a) Electric conductivity $\sigma$ averages the product $\tau v_k^2$ for electrons residing at the chemical potential. As the temperature decreases, $(T_1 < T_2)$ the window of the selected carriers sharpens up. (b) Thermoelectric conductivity $\alpha$ averages the product $\tau v_k$ for electrons residing slightly above and those residing slightly below the chemical potential. The two contributions are subtracted. As the temperature decreases, the window becomes sharper.*

can now deduce expression for $\alpha$ and $\kappa$ reminiscent of the one obtained for $\sigma$:

$$\alpha T = -e \int \tau(k) v_k \cdot v_k (\epsilon - \mu) \frac{\partial f^0}{\partial \epsilon_k} dk \qquad (2.21)$$

$$\kappa = - \int \tau(k) v_k \cdot v_k \frac{(\epsilon - \mu)^2}{T} \frac{\partial f^0}{\partial \epsilon_k} dk \qquad (2.22)$$

The three transport coefficients are expressed in integrals over all $k$-states. These integral contain $\frac{\partial f^0}{\partial \epsilon_k} (\epsilon - \mu)^n$ with an increasing exponent ($n = 0, 1, 2$). Before discussing the physical meaning of all this, let us see that a similar result with more physical transparency can be obtained using the Landauer formalism.

## 2.2.2 The Landauer Formalism

In 1957, Landauer formulated the concept of conduction as transmission. In this approach, the starting point is to consider two electron reservoirs connected together by a wire or any other equivalent solid body, which electrons can use as a path to move from one reservoir to the other. The electric conductance of this object is set by the probability of an electron succeeding in going from one reservoir to the other. Since this probability can take a value between 0 and 1, there should be a finite maximum conductance (and thus a minimum resistance), even if there is no impurity in the way of the travelling electrons.

It took several decades before the relevance of this approach to small structures was recognized. In quantum point contacts, electrons travel ballistically (i.e., without being scattered on their way) from one terminal to the other. It was in these systems, that the quantization of conductance was first observed. An important step was taken by Büttiker [1986] who formulated a solid formalism used to treat electronic transport in mesoscopic structures [Sivan and Imry 1986; Streda 1989; Butcher 1990].

In the case of bulk systems, the Landauer approach yields results which are reminiscent of those traditionally obtained through the Boltzmann approach, but it has the advantage of separating the universal and material-specific aspects of electronic transport in an explicit way.

Consider electrons flowing along a one-dimensional conductor. Let us call $T(\epsilon)$ the probability of a wave-packet with energy $\epsilon$ transmitted across the wire. The charge current is basically the rate of such flow for a quantum of charge $e$. It is obtained by integrating $T(\epsilon)$ over all possible energies, taking into account their probability according to the Fermi-Dirac distribution, $f(T, \epsilon)$:

$$I = \frac{2e}{h} \int f(T, \epsilon) T(\epsilon) d\epsilon \qquad (2.23)$$

The factor two comes from spin degeneracy. The Planck constant provides the appropriate dimensional normalization factor, ensuring a result expressed in proper units.

In the same manner, given the amount of entropy carried by an electron, a similar expression for the flow of heat can be written as

$$Q = \frac{2}{h} \int (\epsilon - \mu) f(T, \epsilon) T(\epsilon) d\epsilon \tag{2.24}$$

Now, the application of a voltage $\Delta V$ and/or a temperature difference $\Delta T$ would lead to a local perturbation of temperature and chemical potential:

$$T = T_0 - \Delta T \tag{2.25}$$

$$\mu = \mu_0 - e\Delta V \tag{2.26}$$

If these deviations are small enough, then

$$f(T, \epsilon) \approx f^0 + \frac{\partial f}{\partial \epsilon} \left[ e\Delta V - \frac{\epsilon - \mu}{T} \Delta T \right] \tag{2.27}$$

where we find again the equilibrium Fermi-Dirac distribution $f^0$.

Now, let us rewrite the pair of equations linking transport coefficients to the flux of heat and charge (Eq. 1.3) adapted to our one-dimensional context:

$$I = G\Delta V - L\Delta T \tag{2.28}$$

$$Q = LT\Delta V - K\Delta T \tag{2.29}$$

Note that, here, the units are simple and scaleless. It is simply ampere for $I$ and watt for $Q$. In one dimension, we don't need to worry about the geometric factor.

Now let us write again Eqs 2.23 and 2.24 using Eq. 2.27. We find

$$I = \frac{2}{h} \int e \frac{\partial f}{\partial \epsilon} \left[ e\Delta V - \frac{(\epsilon - \mu)}{T} \Delta T \right] T(\epsilon) d\epsilon \tag{2.30}$$

$$Q = \frac{2}{h} \int (\epsilon - \mu) \frac{\partial f}{\partial \epsilon} \left[ e\Delta V - \frac{(\epsilon - \mu)}{T} \Delta T \right] T(\epsilon) d\epsilon \tag{2.31}$$

The three transport coefficients, charge conductance ($G$), thermoelectric conductance ($L$), and thermal conductance ($K$), are deduced by comparing Eqs 2.30 and 2.31 with 2.28 and 2.29:

$$G = 2\frac{e^2}{h} \int -\frac{\partial f}{\partial \epsilon} T(\epsilon) d\epsilon \tag{2.32}$$

$$L = 2\frac{e^2}{h} \frac{k_B}{e} \int -\frac{(\epsilon - \mu)}{K_B T} \frac{\partial f}{\partial \epsilon} T(\epsilon) d\epsilon \tag{2.33}$$

$$\frac{K}{T} = 2\frac{e^2}{h} \left(\frac{k_B}{e}\right)^2 \int \left[\frac{(\epsilon - \mu)}{K_B T}\right]^2 \frac{\partial f}{\partial \epsilon} T(\epsilon) d\epsilon \tag{2.34}$$

## 2.3   The Transport Coefficients

The first thing to notice about the three previous equations are the quantized units in which they appear. The quantum of electric conductance is defined by the electron charge and the Planck constant and given by the expression $\frac{e^2}{h}$. It quantifies what a single electronic mode can transmit across a channel when the transmission is ballistic. The quantum of thermal conductance is $\frac{k_B^2}{h}$ and associates Boltzmann and Planck constants. Finally, the three fundamental constants come together to yield the quantum of thermoelectric conductance, $\frac{k_B e}{h}$. Table 2.1 gives their magnitude in SI. units.

Another important aspect is distinguishing what is specific to the material explored and what is due to the transport probe used. The transmission function depends on the details of the electronic structure of the conductor and on the scattering mechanisms that the electrons suffer in the system under study. The transport probe averages a convolution of this function and a kernel, which differs for $\sigma$, $\alpha$, and $\kappa$. This kernel is a product of the energy derivative of the Fermi-Dirac distribution and a power of $(\epsilon - \mu)$.

As seen in Fig. 2.4, this kernel leads to a peculiar selection of relevant carriers for different transport coefficients. In the case of charge conductivity, it probes electrons residing on the Fermi surface. In the case of entropy transport (that is both thermoelectric and thermal), the electrons who contribute are slightly lower or slightly higher than the chemical potential. The profile explains why thermal and electrical conductivities can only be positive, while thermoelectric conductivity can present either of the two signs.

**Table 2.1** *Quanta of the transport coefficients and their magnitude. The variables G, L, and K, are respectively, charge conductance, thermoelectric conductance, and thermal conductance.*

| Transport coefficient | In fundamental constants | Magnitude | SI units |
|---|---|---|---|
| G | $e^2/h$ | $3.87 \times 10^{-5}$ | $\Omega^{-1}$ |
| L | $(k_B e)/h$ | $3.33 \times 10^{-9}$ | $AK^{-1}$ |
| K/T | $k_B^2/h$ | $2.87 \times 10^{-13}$ | $WK^{-2}$ |

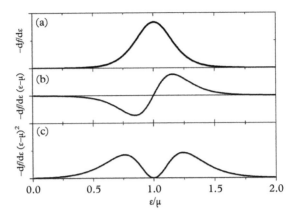

**Figure 2.4** *Kernels of the integrals defining transport coefficients. (a) Electric conductivity, σ. (b) Thermoelectric conductivity α and (c) thermal conductivity κ. In the case of thermoelectric and thermal transport, the probed electrons do not reside at the chemical potential.*

## 2.4   Equivalence of the Two Formalisms

The expressions for the transport coefficients were derived in the previous sections using the Landauer formalism. They have the merit of simplicity and transparency. They allow one to see the units in which each transport coefficient is expressed in and to notice that each transport coefficient, irrespective of the specific structure of the material under study, probes a distinct energy window at the chemical potential or in its neighbourhood. But as mentioned above, the Landauer formalism is based on the analysis of one-dimensional transport between two reservoirs. One may wonder how they connect to the transport properties of a solid body, which was the subject of the Boltzmann formalism. It happens that the two formalisms can be easily translated to each other [Jeong *et al.* 2010]. This can be done by introducing a function $\Xi(\epsilon)$, dubbed transport distribution function [Mahan and Sofo 1996; Scheidemantel *et al.* 2003] and defined as

$$\Xi(\epsilon) = \frac{h}{2}\int \tau(k)v_k \cdot v_k \delta(\epsilon - \epsilon(k))d^3k \tag{2.35}$$

Assuming that energy $\epsilon$ is a smooth function of the wave-vector $k$ and presents no singularity, one can write

$$\Xi(\epsilon) = \frac{h}{2}N(\epsilon)\tau(k)v_k \cdot v_k \tag{2.36}$$

where $N(\epsilon)$ is the density of states

$$N(\epsilon)d\epsilon = d^3k \tag{2.37}$$

It quantifies the link between the number of states per energy and the number of states per wave-vector. In three dimensions, it is proportional to $\epsilon^{1/2}$.

Using the transport distribution function, Eqs 2.16, 2.21, and 2.22 can be written as integrals over energy:

$$\sigma = -2\frac{e^2}{h} \int \frac{\partial f^0}{\partial \epsilon_k} \Xi(\epsilon)\,d\epsilon \tag{2.38}$$

$$\alpha = -2\frac{ek_B}{h} \int \frac{(\epsilon - \mu)}{k_B T}\frac{\partial f^0}{\partial \epsilon_k} \Xi(\epsilon)\,d\epsilon \tag{2.39}$$

$$\kappa/T = 2\frac{k_B^2}{h} \int \left[\frac{(\epsilon - \mu)}{k_B T}\right]^2 \frac{\partial f^0}{\partial \epsilon_k} \Xi(\epsilon)\,d\epsilon \tag{2.40}$$

Comparing these equations with Eqs 2.32–2.34, we see that they are identical, save for the replacement of the function $\mathcal{T}(\epsilon)$ by the function $\Xi(\epsilon)$. Let us comment on this difference. The function $\mathcal{T}(\epsilon)$ is a transmission probability which can take a value between 0 and 1, adapted to the case of a one-dimensional conductor. The function $\Xi(\epsilon)$, on the other hand, is a property of the bulk material under study and is determined by the peculiar $k$-dependence of the velocity and the scattering time as well as the density of states. The latter depends on the dimensionality of the system.

In the following, we will use the bulk expressions for the transport coefficients. One can go back and forth between a bulk sample and its one-dimensional counterpart by replacing $\mathcal{T}(\epsilon)$ with $\Xi(\epsilon)$ and vice versa.

## 2.5 The Wiedemann-Franz Law and the Mott Formula

The Sommerfeld expansion allows one to go one step further and transform the integrals to development series in powers of $\frac{k_B T}{\epsilon_F}$. The first term in the expressions for the three transport coefficients will then become

$$\sigma = 2\frac{e^2}{h}[\Xi(\epsilon_F) + \ldots] \tag{2.41}$$

$$\alpha = 2\frac{\pi^2}{3}\frac{k_B e}{h}k_B T\left[\frac{\partial \Xi(\epsilon_F)}{\partial \epsilon}\Big|_{\epsilon = \epsilon_F} + \ldots\right] \tag{2.42}$$

$$\kappa = 2\frac{\pi^2}{3}\frac{k_B^2}{h}T[\Xi(\epsilon_F) + \ldots] \tag{2.43}$$

These simplified expressions allow one to see two well-known links between the transport coefficients. The first one links thermal and electrical conductances:

$$\kappa = \frac{\pi^2}{3}\frac{k_B^2}{e^2}T\sigma \tag{2.44}$$

In other words, the ratio of the thermal and charge conductances divided by temperature is a universal value called the Lorenz number $L_0$:

$$L_0 = \frac{\kappa}{\sigma T} = \frac{\pi^2}{3} \frac{k_B^2}{e^2} \tag{2.45}$$

The second link is between electric and thermoelectric conductances:

$$\alpha = \frac{\pi^2}{3} \frac{k_B^2}{e} T \frac{\partial \sigma}{\partial \epsilon} |_{\epsilon=\epsilon_F} \tag{2.46}$$

In other words, the thermoelectric conductance is proportional to the energy derivative of the charge conductance. This implies a link between electric conductivity and the Seebeck coefficient. Using Eq. 1.9, one can write

$$S = \frac{\pi^2}{3} \frac{k_B^2}{e} T \frac{\partial ln(\sigma)}{\partial \epsilon} |_{\epsilon=\epsilon_F} \tag{2.47}$$

This equation is often called the Mott formula, as it was first formulated in the thirties by Mott and Jones [1936].

## 2.6   A Physical Picture

In this section, we wish to focus on the physics behind the set equations 2.38–40. We saw that the three transport coefficients are expressed as the product of a quantized physical quantity and the convolution of two functions. Of these two functions, one is the material-specific transport distribution function, and the other is a function, $I_n$, defined as

$$I_n = \left[ \frac{(\epsilon - \mu)}{k_B T} \right]^n \frac{\partial f^0}{\partial \epsilon_k} \tag{2.48}$$

For each transport coefficient, the index $n$ takes one of the three possible values of 0, 1, or 2. The profiles of the three $I_n$ functions are plotted in Fig. 2.4. Let us see what is probed by $\alpha$ in the case of a three-dimensional metal.

The precise energy dependence of transport distribution function $\Xi(\epsilon)$ can be complex in real materials. For the sake of simplicity, let us focus on a one-band isotropic metal with a quadratic dispersion.

In such a system, the density of states increases with an energy as $N(\epsilon) \propto \epsilon^{1/2}$. The energy dependence of the velocity is also $v_k(\epsilon) \propto \epsilon^{1/2}$. As for the energy dependence of the scattering time, we have the choice between two different, both simplistic, approximations. Let us first assume that scattering events occur when electrons, irrespective of their energy, suffer a collision with impurities after travelling an average distance, the so-called mean-free-path. Since the more energetic electrons finish their trajectory faster than the less energetic ones, their scattering time $\tau$ is shorter. In that case, the energy dependence of $\tau$, which would decrease with increasing energy, would attenuate the energy dependence of $N(\epsilon)$ and $v_k(\epsilon)$, and the overall dependence $\Xi(\epsilon)$ would be linear. A second option would be to imagine that, after a given approximate time, a scattering event occurs irrespective of the scattered electron's energy. In this case (i.e. an energy-independent $\tau$), the energy dependence of $\Xi(\epsilon)$ would follow $\epsilon^{3/2}$.

As seen in Fig. 2.5, the outcome would differ somewhat by the choice of one of these approximations, but not qualitatively. The convolution of $I_1$ and $\Xi(\epsilon)$ would subtract a signal by carriers residing somewhat higher than the chemical potential minus those residing somewhat lower than the chemical potential.

If the slope of $\Xi(\epsilon)$ does not significantly change within the energy distance between the negative and positive peaks of $I_1$, then this distance sets the magnitude of the signal. It is proportional to $k_B T$, implying a Seebeck coefficient linear in temperature dependence. If $\Xi(\epsilon)$ is $\epsilon$-linear, this would obviously be the case. If it is $\epsilon^{3/2}$ (i.e. in the case of an energy-independent scattering time), the magnitude of the thermoelectric response would be enhanced but its temperature dependence would not differ.

Now imagine a case in which the scattering time sharply peaks at a given energy near the chemical potential. This happens when there is a specific physical process which can efficiently scatter electrons of a particular energy. In such a case of resonant scattering, the energy dependence of $\tau(\epsilon)$ can be sharp enough to dominate the outcome of $\Xi(\epsilon)$ in a limited energy window. Even in this case, a linear temperature behaviour would be eventually restored when the temperature was lowered further to reach a region where the slope of $\Xi(\epsilon)$ does not change within a window as sharp as $k_B T$. As we will see, in our experimental survey, this appears to be the case of Kondo impurities in metals.

We will come back to the physics behind the thermoelectric response of Fermi liquids in Chapter 5. For the moment, let us end the chapter with a discussion of the sign of the Seebeck coefficient.

## 2.7   Electrons and Holes

An examination of Fig. 2.5 would lead us to the following conclusion. The sign of the Seebeck coefficient is set by the difference between the magnitude $\Xi(\epsilon)$ above and below the chemical potential. The figure seems to suggest that when $\Xi(\epsilon)$ is a monotonous function of energy, this sign should always be the same. However, this is not the complete story, Since the sign of the energy derivative of the Fermi-Dirac distribution is set by the curvature of the Fermi surface.

If the Fermi surface is a full circle or a full sphere, then the unoccupied states (at $T = 0$ limit) are more dense than those which are occupied. This can be seen in Fig. 2.6 by comparing the circumference of the circle lying just outside the Fermi circle. This yields the sign of the Seebeck coefficient, which is negative in this case.

Now, consider a Fermi surface consisting of filled states surrounding a shallow circle or sphere (Fig. 2.6). This situation arises, for example, when the electron concentration becomes high enough to generate a Fermi surface so large that it cannot be contained in a single Brillouin zone [Ashcroft and Mermin 1976]. In this hole-like Fermi surface, the unoccupied states are less dense, as one can see by comparing the circumference of the circle lying just outside the Fermi surface with the Fermi circle.

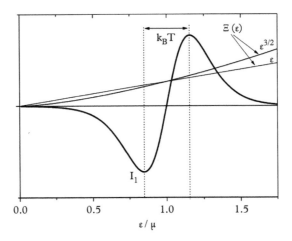

**Figure 2.5** *Thermoelectric response probes electrons above and below the chemical potential. It is set by the convolution of the two function $I_1(\epsilon)$ and $\Xi(\epsilon)$ plotted in this figure. The two expressions for $\Xi(\epsilon)$ correspond to energy-independent mean-free-path and energy-independent relaxation-time approximations.*

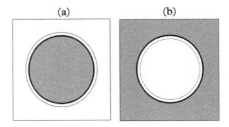

**Figure 2.6** *Electron-like and hole-like Fermi surfaces in a square Brillouin zone. (a) In an electron-like Fermi surface, the density of the occupied states is larger than that of the unoccupied states. (b) In a hole-like Fermi surface, it is the opposite.*

Thus, the sign of the thermoelectric response is generally set by the curvature of the Fermi surface. If it is concave, the unoccupied states are more numerous and, if it is convex, it is the opposite. Thus, the thermoelectric response of holes is positive and that of electrons is negative. The situation becomes more complicated when the Fermi surface presents different local curvatures at different positions. Note also that a pathological energy dependence of the scattering time can also complicate the final outcome. We will return to this issue in a discussion on the comparative signs of Seebeck and Hall effects in the next chapter.

# 3

# Non-Diffusive Thermoelectricity

## 3.1 Electrons and Phonons

In the previous chapters, we considered how electrons respond to a thermal gradient and/or electric field. We did not take into account the fact that electrons in a solid are not alone. They are surrounded by potentially vibrating atoms. The electron gas coexists with a phonon gas. Phonons play a major role in transport properties of solids. Their presence is felt differently when one probes electric, thermal, or thermoelectric transport.

Phonons do not carry electric charge and therefore do not contribute to electric conductivity. In this case, they are only invoked as scattering centres. By exchanging energy and impulsion with charge-carrying electrons, they may impede charge conductivity. However, there is no such thing as a lattice contribution to charge transport.

On the other hand, since phonons do carry energy, they do contribute to thermal conductivity. In a non-insulating solid hosting phonons and mobile electrons, any discussion of thermal conductivity is complicated by the fact that both electrons and phonons carry heat and scatter each other, leading to non-monotonous variations with temperature and/or carrier concentration.

In the case of thermoelectricity, phonons play a more subtle role. Since they are not charged, they do not directly respond to an electric field and their flow does not generate an electric field. Nonetheless, in the years following the second world war, on both sides of the iron curtain it was discovered that there is such thing as a phonon contribution to thermoelectricity. The expression 'phonon drag' was coined to describe this contribution, arising from the fact that phonons and electrons pull each other.

## 3.2 Phonon Drag: The Basic Picture

Let us begin by considering the ultimate reason behind this phenomenon. Consider an experiment to measure the Seebeck coefficient of a solid: in which a thermal gradient is imposed and the electric field generated is measured, while the charge flow is kept equal to zero. In the presence of a temperature gradient, crystal vibrations tend to drift from the hot side to the cold side and the phonon population is not homogenous along the sample.

*Fundamentals of Thermoelectricity*. First Edition. Kamran Behnia.
© Kamran Behnia 2015. Published in 2015 by Oxford University Press.

But this means that charge carriers travelling along the thermal gradient do not suffer the same rate of scattering events as those travelling in the opposite direction. This would generate a net *charge* current along the thermal gradient. However, the experimental set-up is designed in such a way that it is impossible to keep the flow of charge going. The only way for the system to meet this situation is to build up a finite electric field to compensate for the excess scattering the charge carriers suffer when they travel along one of the two orientations. This additional electric field will be on top of the original diffusive response of the electrons and constitutes the phonon-drag contribution to the Seebeck coefficient.

In an equivalent picture, one may consider a Peltier set-up. In this case, the experiment is designed to measure the ratio of energy flow to the charge flow, in the total absence of any temperature difference along the sample. There is now a net flow of electrons along the sample. The scattering rate between electrons and phonons travelling along each other is not identical to those moving in opposite directions. This generates an imbalance in the phonon population travelling in opposite directions and, as a consequence, a net flow of energy in addition to the purely electronic one. This additional flow of energy is the phonon-drag contribution to the Peltier effect.

L. E. Gurevich is known to be the founding father of this concept, which was discovered in the early 1950s by western physicists, [Frederikse 1953; Herring 1954; Macdonald 1954; Herring 1958].

Two distinct sets of experimental observations provided the motivation to go beyond diffusive thermoelectricity and look more carefully at phonons and the significant role they play in the thermoelectric response of electrons. The first observation was the sharp increase in the magnitude of the thermoelectric response in elemental semiconductors such as germanium and silicon below room temperature [Frederikse 1953; Geballe and Hull 1954, 1955]. The second was the fact that the Seebeck coefficients of noble metals (copper, silver, and gold) displayed intriguing humps in their temperature dependence. Soon, a consensus emerged that phonon drag is the origin of both phenomena.

In the following decades, several theories for phonon drag were proposed. They fall into two categories. The first set, elaborated by Macdonald [1962], Bailyn [1958, 1960, 1967], Ziman [1960], and Guénault [1971] is a theory for electrons in metals and quantifies the change in the electron flow caused by the presence of phonons. The second, formulated by Herring [1954] for electrons in semiconductors, quantifies the force on phonons due to the flow of electrons. A detailed review of phonon drag in semiconductors can be found in a paper by Gurevich and Mashkevich [1989]. In the following sections, we are going to review both sets of theories and various experimental attempts to check their relevance.

## 3.3   Phonon Drag in Metals

The picture proposed by Macdonald [1962] permits us to evaluate the magnitude of the phonon-drag contribution to the Seebeck coefficient. Imagine a gas of phonons in which electrons are travelling. The average pressure, the electrons feel would be

$$P = \frac{1}{3}U(T) \tag{3.1}$$

Here, $U$ is the energy per unit of volume in this 'gas', which depends on absolute energy. When the temperature is homogenous, this pressure is isotropic and has no effect on any preferential orientation for electron travel. But a temperature gradient would generate a pressure gradient; in other words, a net force on electrons along the temperature gradient. The density of this force would be

$$f = -\nabla P = -\frac{1}{3}\frac{U}{T}\nabla T \tag{3.2}$$

As mentioned above, this force is to be counterbalanced to keep the net flow zero. This can be done by an electric field acting on the charge of carriers, with a density of

$$f = neE \tag{3.3}$$

Here, $n$ is the carrier density, and $e$ is the charge of electron. Now, the temperature derivative of the phonon's energy density, $U(T)$, is basically the lattice specific heat $C_L$. So, all this will add up to

$$neE = -\frac{1}{3}C_L\nabla T \tag{3.4}$$

In other words, the additional Seebeck signal caused by phonon drag is

$$S_g = \frac{-E}{\nabla T} = -\frac{1}{3}\frac{C_L}{ne} \tag{3.5}$$

The phonon drag is, thus, proportional to the lattice specific heat. This is understandable. The larger the capacity of the lattice to stock energy, the more efficient are phonons to damp the impulsion of the travelling electrons. Phonon drag is also inversely proportional to the carrier density: the lower their concentration, the higher the ability of phonons to pull them.

In the simple picture sketched above, phonons are uniquely scattered by electrons. This is obviously not the case in the real world. Phonons can also be scattered by other phonons and by impurities. Therefore, this picture is only realistic when most of the scattering events the phonons suffer are caused by electrons.

## 3.4   Scattering Phonons

How does the phonon drag evolve with temperature? Let us try to address this question with the simple picture sketched above as our only baggage.

At room temperature and above, in most solids, phonons are most often scattered by other phonons. Such a phonon–phonon scattering event requires at least one phonon among those involved in scattering to be energetic enough to provide a wave-vector as large as the unit vector of the reciprocal lattice. As the temperature decreases, such high-energy phonons become rare. This strongly reduces the frequency of phonon–phonon scattering events. As a consequence, phonon-electron scattering events become more frequent at low temperatures, generating significant phonon-drag thermoelectricity. At still lower temperatures, the phonon energy decreases further and the typical phonon wavelength becomes much longer than the wavelength of mobile electrons (that is $\lambda_F$, the Fermi wavelength). In this situation, phonons are too extended to be scattered by electrons and again no significant phonon drag is expected. Thus, roughly speaking, the phonon drag is expected to peak at an intermediate temperature well below room temperature, but significantly above one Kelvin. Let us put this in more detail.

Phonons are quasi-particles with a momentum $\hbar q$ and an energy $\hbar \omega_q$. As the temperature is changed, their population follows Bose-Einstein statistics.

$$N_q = \frac{1}{exp(\hbar \omega_q / k_B T) - 1} \tag{3.6}$$

In the case of acoustic phonons, there is a linear dispersion relation between the frequency $\omega$ and the wave-vector $k$ for small wave-vectors:

$$\omega_q = v_s q \tag{3.7}$$

The sound velocity $v_s$ depends on orientation and differs among different vibrational modes. Like mobile electrons, phonons are carriers of energy and momentum. But, there is an important difference. In the case of electrons, only a small fraction, those with an energy very close to the Fermi energy, participate in transport. This remains true whatever the temperature. As the sample is cooled down, the typical heat-carrying electron has its energy $\epsilon_F$ and its wave-vector $k_F$ within a finite window, which becomes narrower as the temperature decreases.

On the other hand and unlike electrons, the energy of heat-carrying phonons is not locked to a temperature-independent value. The typical phonon energy increases as the system is warmed up. They travel at the same velocity at all temperatures. However, as their typical energy becomes larger with increasing temperature, their typical wavelength decreases. The energy scale of the phonon system is the Debye temperature, $\Theta_D$, which is the energy required to excite phonons with highest frequency (and thus lowest wavelength) possible. This is of course set by atomic separation, $a_d$. Therefore, one can roughly link the three material-dependent quantities:

$$\Theta_D \approx \hbar v_s / a_d \tag{3.8}$$

Since the typical wavelength of phonons varies with temperature, their exposure to scattering centres is different in different temperature windows.

Let us begin with high temperatures. In the presence of anharmonic coupling between the vibrational modes, the notion of distinct quasi-particles can be preserved by invoking phonon–phonon scattering. During such an event, energy and impulsion would be redistributed between incoming and outgoing particles. In a normal process, the momentum is conserved. But, there are scattering events called Umklapp, in which there is a difference in the total momentum before and after the scattering event, which is an integer multiple of the elementary wave-vector in the reciprocal space. These scattering events are the reason why, even in absence of impurities and mobile electrons, a perfect electric insulator is expected to display a finite thermal resistance.

Above the Debye temperature, Umklapp scattering between phonons is the main scattering event, and phonon drag is not significant. As the temperature decreases, phonons begin to feel the presence of mobile electrons. Electrons scatter phonons more efficiently when the typical phonon wavelength is comparable to the Fermi wavelength. In this temperature range, usually centred on a temperature which is a fraction of $\Theta_D$, phonon drag peaks. At still lower temperatures, one expects that, because of their long spatial extension, electrons become invisible to phonons. However, crystal defects continue to scatter them. The typical phonon wavelength continues to grow with decreasing temperature until it attains crystal dimensions. In this situation, phonons travel ballistically from one end of the crystal to the other.

Thus, it is obvious that Eq. 3.5 is not complete and needs at least one other factor which quantifies the proportion of scattering events occurring to phonons and implying electrons. Various attempts to remedy this have been reported.

## 3.5   Anisotropic Phonon Drag

An expression derived by Bailyn [1967] has the merit of finding that phonon drag can be very anisotropic and can be of both signs. Here, we will present the simple derivation of this expression presented by Guénault [1971].

The question to be addressed is the following: What is the electric current density generated by a thermal gradient along the sample and due to phonon drag?

Let us imagine that, at each instant, a number of scattering events occurs with the participation of a phonon with a wave-vector $q$. These scattering events alter the mean-free-path of the electrons involved. Guénault argued that the charge current density caused by a thermal gradient along the $x$-axis through this process can be written as

$$\mathcal{J}_g^x = \sum_q -\left(\frac{\partial N_q}{\partial t}\right)_{ph-e} e < \ell_{k'} - \ell_k >_x \tag{3.9}$$

The first term in the right side of this equation is the rate of these events. Here, $N_q$ is the density of phonons with a wave-vector $q$. The relation $\left(\frac{\partial N_q}{\partial t}\right)_{ph-e}$ represents the total number of phonons per unit volume and per unit time, leaving state $q$ as a consequence of electron–phonon collisions. Summing over all $q$s will give the overall rate. The product of the two other terms yields the incremental change in current density caused by each of these events: a charge of $e$ is displaced by the difference in the incoming and outgoing mean-free-paths. As we are interested in the current parallel to the thermal gradient, the average is taken along the orientation of the thermal gradient.

It is fair to assume that, for each wave-vector $q$, phonon–electron collisions are only a fraction $\alpha(q)$ of the total collisions:

$$\left(\frac{\partial N_q}{\partial t}\right)_{ph-e} = \alpha(q)\left(\frac{\partial N_q}{\partial t}\right)_{col.} \tag{3.10}$$

Another plausible assumption is that the phonon number at a given $q$ is conserved. The change in their number caused by collisions is counterbalanced by the drift induced by the temperature gradient. In other words

$$\left(\frac{\partial N_q}{\partial t}\right)_{col.} + \left(\frac{\partial N_q}{\partial t}\right)_{drift} = 0 \tag{3.11}$$

But, the drift-induced change is proportional to the thermal gradient:

$$\left(\frac{\partial N_q}{\partial t}\right)_{drift} = -\frac{\partial N_q^0}{\partial T}\frac{dT}{dx}v_{qx} \tag{3.12}$$

Here, $v_{qx}$ is the velocity of a phonon with a wave-vector $q$ along the $x$-axis. Finally, in the relaxation-time approximation, the mean-free-path is related to scattering time through the Fermi velocity:

$$\ell_k = v_k \tau_k \tag{3.13}$$

Eq. 3.9 can now be rewritten as

$$\mathcal{J}_g^x = \sum_q -\left(\frac{\partial N_q^0}{\partial T}\right)\frac{dT}{dx}v_{qx}\alpha(q)e < v_{k'}\tau_{k'} - v_k\tau_k >_x \tag{3.14}$$

The phonon-drag Seebeck coefficient $S_g$ is equal to the electric field caused by this current density divided by the temperature gradient. Therefore,

$$S_g^x = \frac{e}{\sigma}\sum_q -\left(\frac{\partial N_q^0}{\partial T}\right)v_{qx}\alpha(q) < v_{k'}\tau_{k'} - v_k\tau_k >_x \tag{3.15}$$

and, assuming the cubic symmetry,

$$S_g = \frac{1}{3}\frac{e}{\sigma}\sum_q -\left(\frac{\partial N_q^0}{\partial T}\right)\overline{v_q}\alpha(q) < v_{k'}\tau_{k'} - v_k\tau_k > \tag{3.16}$$

In the simple case of free electrons with $\alpha(q) = 1$ and isotropic scattering time,

$$S_g = \frac{1}{3} \frac{1}{n\mu} \tau \sum_q - \left( \frac{\partial N_q^0}{\partial T} \right) \overline{v_q} < v_{k'} - v_k > \qquad (3.17)$$

For acoustic phonons, velocity is a linear function of the wave-vector. Moreover, for free electrons, the difference in velocity before and after the collision is set by the effective mass and the change in the wave-vector, further simplifying the expression:

$$S_g = \frac{1}{3} \frac{m^*}{ne} \sum_q - \left( \frac{\partial N_q^0}{\partial T} \right) \frac{\omega_q}{q} \frac{\hbar q}{m^*} \qquad (3.18)$$

Finally,

$$S_g = \frac{1}{3ne} \sum_q - \left( \frac{\partial N_q^0}{\partial T} \right) \hbar \omega_q \qquad (3.19)$$

We can now see that, in these conditions, the equation becomes identical to Eq. 3.5 originally derived by Macdonald [1962].

More generally, Eq. 3.15 leads to a number of expectations. First of all, the temperature dependence of the effect is not only set by the lattice specific heat, which quantifies the number of phonons available to leave state $q$, but also by the fraction of phonons in state $q$ scattered by electrons, $\alpha(q)$. The first term leads to a $T^3$ behaviour for $T \ll \Theta_D$ and becomes temperature independent at $T \geq \Theta_D$. The temperature dependence of the second term is more complicated. It peaks when phonon–electron scattering is the dominant mechanism for phonon scattering. At high temperatures, $1/T$ temperature dependence of the phonon–phonon scattering time is expected to set the tendency.

Secondly, the phonon drag can be anisotropic. When the electronic mean-free-path is anisotropic, the last term in the right side of the equation can be different according to the orientation along which the thermal gradient is applied.

Finally, the phonon-drag contribution to the Seebeck coefficient is not necessarily negative. In a complicated Fermi surface consisting of components with different curvatures, the last term can be positive or negative for different combinations of $k$ and $k'$ leading to an indeterminate sign for the final sum.

## 3.6 Phonon Drag in Semiconductors

Long before Bailyn, but with germanium and silicon (and not metals) in mind, Herring proposed a quantitative estimation of phonon-drag Seebeck effect [Herring 1954]. The motivation was the availability of extensive thermoelectric data on these systems [Frederikse 1953; Geballe and Hull 1954, 1955], data which pointed to a puzzlingly large Seebeck coefficient at low temperature (See Chapter 6).

In this picture, phonon drag is quantified from the phonon point of view. Herring equated the rate which the phonon system receives crystal momentum from electrons to the rate it loses it through Umklapp collisions and boundary scattering :

$$\frac{d\mathbf{P}}{dt}\Big|_e = f^{el.} ne\mathbf{E} \tag{3.20}$$

Here, $\mathbf{P}$ is the crystal momentum per unit volume and $f^{el.}$ is the fraction of the momentum lost by electrons to the lattice. In absence of impurity scattering of electrons, $f^{el.} = 1$. The force per unit volume is $ne\mathbf{E}$, where $\mathbf{E}$ is the electric field, and $n$ is the carrier density. Now, one may define a time scale for the relaxation of the phonon system, that is,

$$\frac{d\mathbf{P}}{dt}\Big|_e = \frac{\mathbf{P}}{\tau_{ph-e}} \tag{3.21}$$

The energy flow density $\mathcal{J}^Q$ due to crystal momentum is

$$\mathcal{J}^Q = \mathbf{P}\overline{v}_s^2 \tag{3.22}$$

Here, $\overline{v}_s$ is the average sound velocity. Therefore,

$$\mathcal{J}^Q = \overline{v}_s^2 \tau_{ph-e} f^{el.} ne\mathbf{E} \tag{3.23}$$

One can replace the electric field by charge current density, using $\mathbf{J}^e = ne\mu\mathbf{E}$, and obtain

$$\mathcal{J}^Q = \overline{v}_s^2 \frac{\tau_{ph-e} f^{el.}}{\mu} \mathcal{J}^e \tag{3.24}$$

This leads to an estimation of the phonon-drag contribution to the Peltier coefficient:

$$\Pi_g = \frac{\overline{v}_s^2 \tau_{ph-e}}{\mu} f^{el.} \tag{3.25}$$

Since the Kelvin relation links the Peltier and Seebeck coefficients, the resulting phonon drag thermopower is

$$S_g = \frac{1}{T} \frac{\overline{v}_s^2 \tau_{ph-e}}{\mu} f^{el.} \tag{3.26}$$

Now defining a phonon mean-free-path, $\ell_{ph} = \overline{v}_s \tau_{ph}$, the expression can be written as

$$S_g = \frac{\overline{v}_s}{T} \frac{\ell_{ph}}{\mu} f^{el.} \tag{3.27}$$

This expression sets the expected temperature dependence of the phonon-drag contribution to the thermoelectric response. As the system is cooled down, the phonon mean-free-path increases, amplifying the basic tendency towards the enhancement set by the $1/T$ factor. On the other hand, electron mobility often increases, and the fraction of electrons scattered by phonons, $f^{el.}$, decreases with cooling. The latter two counterbalance the tendency set by the first two. The effect is expected to vanish in the zero-temperature limit, as no electron would give its momentum to phonons.

The three key material-dependent properties here are phonon mean-free-path, electron mobility, and electron–phonon coupling.

## 3.7  Two Competing Pictures

As stated before, historically two brands of phonon drag theory have emerged. One of them was elaborated by Herring for semiconductors by quantifying the Peltier effect due to the phonon energy flow responding to an electronic charge flow. The other was elaborated by Macdonald, Bailyn, Ziman, and Guénault for metals. The latter authors started by quantifying the way force was generated by phonons on electrons in the presence of a thermal gradient. Even if the contexts of their investigations differed (metals vs semiconductors), one would expect that their conclusions converge, as both treat the interaction between electron and phonon gases. A rapid examination of Eq. 3.17 and 3.27 establishes that this is not the case.

In the Herring picture, if phonons were the unique scatterers of electrons, the phonon drag would be proportional to the ratio of the sound velocity to the electron mobility times the phonon mean-free-path divided by temperature:

$$S_g \approx \frac{\overline{v_s}}{\mu} \frac{1}{T} \ell_{ph} \tag{3.28}$$

In order to compare the two pictures, Eq. 3.17 can be rewritten as

$$S_g \approx \frac{\overline{v_s}}{\mu} \frac{1}{T} \frac{<N_{ph}>}{n_e} < \delta \ell_e > \tag{3.29}$$

To simplify the comparison, we have assumed that phonons are uniquely scattered by electrons and that their population smoothly increases with temperature. Two striking differences are apparent. In the first case, phonon drag is proportional to the phonon-mean-free-path. In the second, it is proportional to the average change in the electron-mean-free-path inflicted by phonon scattering. Moreover, in the second expression, the effect is proportional to the ratio of phonon-to-electron concentration. No such pondering factor is present in the first expression.

Surprisingly, both expressions have been used by experimentalists for decades, to quantify the magnitude of phonon drag in various systems. But this significant difference between the two pictures leads to conflicting quantitative predictions and has not been commented upon by either community. Without entering into a detailed analysis of Herring's expression, Sondheimer briefly commented that the expression violates the

Kelvin relation [Sondheimer 1956a]. This early remark has been the only published theoretical comment on this issue known to the author of these lines.

One shall not forget that both pictures sketched above contain numerous simplifications. In one case the electron–phonon coupling is seen through phonons perspective: their flow carrying energy is a simple response to their collisions with electrons. In the other, the same is observed from the electronic point of view: the charge flow arises because of electrons' collision with phonons. A satisfactory treatment of the phonon drag problem needs a self-consistent treatment of interaction reservoirs of phonons and electrons, one obeying Fermi-Dirac and the other, Bose-Einstein distributions. This is what Gurevich and Mashkevich have attempted in their review paper [Gurevich and Mashkevich 1989]. One of their remarkable conclusions is the possible violation of Onsager reciprocal relations in presence of mutual drag between electrons and phonons.

## 3.8   Experimental Evidence for Phonon Drag

How does the phonon drag theory match with experimental facts? As mentioned above, there were two motivations to look for any eventual role of the lattice: the hump in the thermopower of simple monovalent (alkali and noble) metals and the sharp enhancement in the Seebeck coefficient of elemental semiconductors. The phonon drag picture provides a qualitative explanation for both observations. However, quantitative verification of the theory has proven to be difficult. In the case of alkali metals at low temperatures, Hanna and Sondheimer [1957] and Macdonald [1962] have shown that their theory successfully fits the data with a diffusive and phonon-drag term along the lines of Eq. 3.5.

One obvious difficulty in confronting theory with experimental data is the parameter quantifying the fraction of phonons which collide with electrons ($\alpha(q)$ in Eq. 3.15). The magnitude and temperature dependence of this quantity cannot be calculated for a given Fermi surface from first principles. One can only put forward a reasonable guess, and this significantly enhances the fuzziness of any phonon drag analysis. In the case of noble metals, attempts have been reported to explain why the phonon-drag peak in thermopower is positive [Dugdale and Bailyn 1967]. According to a number of investigators [Blatt 1960; Blatt, Garber, and Scott 1964; Huebener 1966], in copper and gold, phonon drag scales with the lattice thermal conductivity rather than with the lattice specific heat. This is compatible with the theory elaborated by Hanna and Sondheimer [1957], which concludes that lattice thermal conductivity and phonon drag are closely linked.

Finally, a theoretical study by Nielsen in 1970 led to the conclusion that, by taking into account second-order scattering processes between electrons and virtual phonons, even the diffusive thermoelectricity can display features that look like the phonon drag contribution to the Seebeck coefficient. Since these second-order processes are particularly important when the temperature is a fraction of the Debye temperature, the relevance of phonon drag as the definitive interpretation of enhanced thermoelectricity was put into question by these authors [Nielsen and Taylor 1970, 1974].

To sum up,the existence of phonon drag is generally accepted among condensed-matter physicists, and the concept of out-of-equilibrium phonons interacting with electrons seems firmly established. On the other hand, in spite of several decades of effort, there are only few cases where the experimental data are quantitatively explained by available theories. The only case where theory appears to be quantitatively successful is the Seebeck coefficient in sodium [Hanna and Sondheimer 1957] and in potassium alloys [Macdonald 1962]. In both cases, both the magnitude of the Seebeck coefficient and its curvature are negative. The positive hump observed in noble metals and in lithium remains controversial. We will come back to experimental data in chapter 6.

Phonon drag is believed to become even more significant at the surface of metals and semiconductors. The drag of surface electrons by non-equilibrium phonons has been directly monitored by studies of acoustoelectric effects [Zavaritsky 1987].

## 3.9  Magnon Drag

Magnons, which are sometimes called spin waves, are the elementary vibrations of a magnetic sublattice. They are present when the solid is magnetically ordered. Their properties are similar to phonons, with a spectrum specific to the magnetic sublattice. The existence of magnon drag has been proposed for reasons similar to phonon drag. Its existence has been reported in iron [Blatt *et al.* 1967] as well as in chromium [Trego and Mackintosh 1968]. A detailed theory for this effect has been elaborated by Bailyn [1962].

# 4

# Magnetothermoelectricity

In the presence of a magnetic field, the three transport coefficients, electric, thermal, and thermoelectric conductivities, reveal their tensorial nature. Take charge transport, for example. Electric conductivity would be no more than just a scalar defining the ratio of electric field to electric current. It would be no longer necessary for these two vectors to be parallel to each other. A similar feature occurs in the case of thermal and thermoelectric phenomena.

It is true that, in anisotropic media, the non-scalar nature of conductivity tensors is observable even in the absence of a magnetic field. Consider an anisotropic conductor, one which conducts electricity better along one axis (say $x$) than the other (say $y$). In this case, if an electric field is applied along an arbitrary orientation (neither $x$ or $y$), it would generate a tilted charge current.

The application of a magnetic field, however, introduces a new possibility. Now, a charge current *perpendicular* to the applied electric field can be induced by the magnetic field. This is the Hall effect. The Nernst-Ettingshausen effect and Righi-Leduc effects correspond to thermoelectric and thermal analogues of the Hall effect. Thermomagnetics or magnetothermoelectricity refer to physical phenomena arising due to the entanglement between these three independent vectors.

In this chapter, we will discuss the tensorial aspect of thermoelectricity induced by the application of a magnetic field. We will limit ourselves to the case of a magnetic field weak enough to remain in the semi-classic limit described by the Boltzmann equation, and we will follow along the line traced by Ziman [1960]. Purely quantum effects such as the Landau quantization appear in strong quantizing magnetic fields and will not be considered here. The reader can find a brief discussion of thermoelectricity in strong magnetic fields in chapter 10.

## 4.1  In the Presence of a Magnetic Field

The Boltzmann equation needs to be amended in presence of a finite magnetic field. The magnetic field generates a Lorenz force proportional to $ev \times B$. This force will affect the electron momentum $\hbar k$. Therefore,

*Fundamentals of Thermoelectricity*. First Edition. Kamran Behnia.
© Kamran Behnia 2015. Published in 2015 by Oxford University Press.

$$\hbar \frac{dk}{dt} = ev_k \times B \tag{4.1}$$

This should affect the transport properties of the system. The paradigm of the Boltzmann formalism brings us to ask the following question: what is the probability of finding an electron with a given wave-vector $k$ at a given position $r$, $f_k(r)$, and how is this probability is affected by the application of a magnetic field?

The way the Lorentz force affects each electron depends on its wave-vector. Therefore, the change in $f_k(r)$ induced by a magnetic field depends on the way the distribution depends on the wave-vector:

$$\frac{df_k(\mathbf{r})}{dt}\bigg|_{Mag\cdot field} = \frac{e}{\hbar}(v_k \times B)\frac{\partial f_k(\mathbf{r})}{\partial \mathbf{k}} \tag{4.2}$$

In plain words, the probability of the presence of such an electron in a given place will change with a rate which is proportional both to the strength of this force and to the momentum derivative of the distribution probability. This is just a consequence of Newton's law, which equates the force with the rate of change of the momentum.

The reader may recall that the electric field affects the distribution in a similar fashion. However, there is a big difference. Suppose that the actual distribution is only barely different from the original (i.e. the equilibrium Fermi-Dirac) distribution, $f_k^0(r)$, so that one can write

$$f_k(r) = f_k^0(r) + g_k(r) \tag{4.3}$$

The function $g_k(r)$ quantifies the deviation from equilibrium. The importance of $g_k(r)$ is very different for electric and magnetic fields. The electric field affects the distribution, even if $g_k(r)$ is negligible, but the magnetic field does not. Indeed, in the presence of a magnetic field

$$\frac{e}{\hbar}(v_k \times B) \cdot \nabla_k f_k^0(\mathbf{r}) = 0 \tag{4.4}$$

This can be seen by rewriting the equation as

$$e(v_k \times B) \cdot v_k \frac{\partial f_k^0(\mathbf{r})}{\partial \epsilon_k} = 0 \tag{4.5}$$

Here, we have used the fact that each electronic state in a Fermi-Dirac distribution has a well-defined group velocity:

$$\hbar v_k = \frac{\partial \epsilon_k}{\partial k} \tag{4.6}$$

Let us briefly pause and consider this. We saw that an electric field rigidly shifts the Fermi-Dirac distribution. A temperature gradient, on the other hand, sharpens one

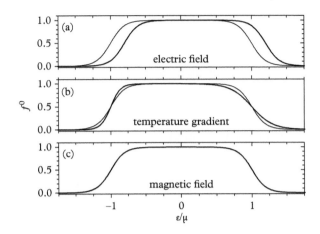

**Figure 4.1** *Magnetic field does not perturb the Fermi-Dirac distribution. In contrast to an electric field (a) and to a temperature gradient (b), the magnetic field does not affect the Fermi-Dirac distribution (c). Its presence is felt only by those electrons which have been already displaced from equilibrium by an electric field or a temperature gradient.*

side of the distribution and broadens the other side. This means that an electric field or a thermal gradient, both of them alone and without any help from the other, can generate a flow of electrons. These are familiar to us as electric conductivity and thermal conductivity. On the other hand, a lonely magnetic field does not disturb the Fermi-Dirac distribution (see Fig. 4.1).[1] There is no purely magnetic analogue of thermal and electric conductivity. *Only those electrons already kicked out of equilibrium by an electric field or a thermal gradient can feel the effect of a magnetic field.*

## 4.2 Magnetoresistance and the Hall Effect

The linearized Boltzmann equation (i.e. Eq. 2.19) in the presence of a magnetic field but in the absence of a thermal gradient becomes

$$e\frac{\partial f_k^0}{\partial \epsilon_{\mathbf{k}}} v_k \cdot E + \frac{g_k(r)}{\tau} + \frac{e}{\hbar}(v_k \times B) \cdot \nabla_k(g_k(r)) = 0 \tag{4.7}$$

In the absence of a magnetic field, the response of the system is set by the relaxation time and the magnitude of the deviation-from-equilibrium function $g_k(r)$. Now, with a magnetic field present, there is another component to the story: the way $g_k(r)$ depends on the $k$. At this stage, let us introduce the notion of electron mobility, $\bar{\mu}$:

---

[1] This is only true when the Landau quantization is negligible.

$$\bar{\mu} = \frac{e\tau}{\hbar}\frac{\partial v}{\partial k} \tag{4.8}$$

Now Eq. 4.7 can be written in the following fashion:

$$\tau e \frac{\partial f_k^0}{\partial \epsilon_k} v_k \cdot E + [1 + \bar{\mu}(v_k \times \mathbf{B}) \cdot \mathbf{V}_v] g_k = 0 \tag{4.9}$$

For the sake of simplification, let us assume for the time being, that mobility $\mu$ is a scalar. With this assumption and using the invariance of the mixed vector under a circular shift, Eq. 4.9 can be rewritten as

$$g_k = \mathbf{v_k} \cdot \left[ \tau e \left( \frac{-\partial f_k^0}{\partial \epsilon_k} \right) \mathbf{E} + \mu \mathbf{B} \times \mathbf{V}_v(g_k) \right] \tag{4.10}$$

This function represents the distribution of electrons kicked out of equilibrium in $k$-space. The first term of the right-hand side of this equation represents what is solely caused by the electric field. Figure 4.2 sketches the angular distribution of these states. There is an excess of electrons travelling along the electric field and a deficit of those going in the opposite direction. The electrons which happen to travel perpendicular to the electric field are not affected.

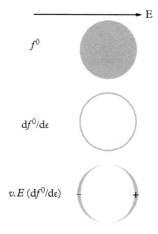

**Figure 4.2** *Electric fields kick electrons out of equilibrium in reciprocal space. Top: The equilibrium distribution implies that electrons can have different velocities up to the Fermi velocity. Middle: Only electrons whose velocity is equal to Fermi velocity participate in transport. Bottom: When the electric field is switched on, an out-of-equilibrium population of electrons is generated. The difference in electron density for a given velocity is positive for those travelling along the electric field negative among those travelling in the opposite direction and nil for those travelling perpendicular to the electric field.*

The second term of the right-hand side of Eq. 4.10 is the vector product of two vectors, the magnetic field and the velocity gradient of the out-of-equilibrium population. As a possible solution for this equation, let us try a trial function such as

$$g_k = \mathbf{v_k} \cdot \mathbf{G} \tag{4.11}$$

When there is no magnetic field, the orientation of vector $\mathbf{G}$ is set by the electric field. Its magnitude in such a case, $\mathbf{G_0}$, would be

$$\mathbf{G_0} = \tau e \frac{-\partial f_k^0}{\partial \epsilon_k} \mathbf{E} \tag{4.12}$$

This represents the component of the out-of-equilibrium distribution (before its selective amplification by velocity) in the absence of a magnetic field. We need now to estimate $\mathbf{G}$ in the presence of a magnetic field. Let us assume that $\mathbf{G}$ is independent of velocity. Then, combining Eqs 4.10, 4.11, and 4.12, one obtains

$$\mathbf{G} = \mathbf{G_0} + \mu\mathbf{B} \times \mathbf{G} \tag{4.13}$$

These vectors are sketched in Fig. 4.3. This equation can be solved by a simple trick. Let us multiply both sides by an identical term:

$$\mathbf{G} \times \mu\mathbf{B} = [\mathbf{G_0} + \mu\mathbf{B} \times \mathbf{G}] \times \mu\mathbf{B} \tag{4.14}$$

Replacing the left-hand side of Eq. 4.14 using Eq. 4.13, one can write

$$\mathbf{G_0} - \mathbf{G} = \mathbf{G_0} \times \mu\mathbf{B} + [\mu\mathbf{B} \times \mathbf{G}] \times \mu\mathbf{B} \tag{4.15}$$

Using a simple rule of vector algebra, this becomes

$$\mathbf{G_0} - \mathbf{G} = \mathbf{G_0} \times \mu\mathbf{B} + \mu^2 B^2 \mathbf{G} - (\mu^2 \mathbf{B} \cdot \mathbf{G})\mathbf{B} \tag{4.16}$$

$$\mathbf{G} = \frac{\mathbf{G_0} - \mathbf{G_0} \times \mu\mathbf{B} + (\mu^2 \mathbf{B} \cdot \mathbf{G})\mathbf{B}}{1 + \mu^2 B^2} \tag{4.17}$$

Finally, taking into account Eq. 4.13, this last equation simplifies to

$$\mathbf{G} = \frac{\mathbf{G_0} - \mathbf{G_0} \times \mu\mathbf{B} + (\mu^2 \mathbf{B} \cdot \mathbf{G_0})\mathbf{B}}{1 + \mu^2 B^2} \tag{4.18}$$

This leads to a satisfactory solution for $g_k$:

$$g_k = -\tau e \frac{\partial f^0}{\partial \epsilon_k} \mathbf{v_k} \cdot \frac{\mathbf{E} + \mu\mathbf{B} \times \mathbf{E} + (\mu^2 \mathbf{B} \cdot \mathbf{E})\mathbf{B}}{1 + \mu^2 B^2} \tag{4.19}$$

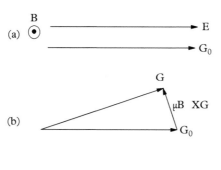

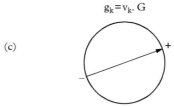

**Figure 4.3** *The effect of magnetic field on out-of-equilibrium electrons. (a) The vector $G_0$ represents the out-of-equilibrium depopulation of electrons, as illustrated in the bottom panel of the previous figure. In the absence of a magnetic field, it is parallel to the situation. (b) In the presence of a field perpendicular to the electric field, it is replaced by the vector $G$, which should satisfy Eq. 4.13. (c) The population of electrons which contribute most positively and negatively to the overall charge current has now a skewed distribution.*

Thus, the presence of a magnetic field modifies the out-of-equilibrium population in three different ways. First, the component parallel to the electric field is reduced. This reduction is set by the magnitude of the denominator. In addition to this original component, two new components emerge. The second component is parallel to the vector product of magnetic field and electric field (the second term in the nominator). Finally, there is a third term in the nominator, which is *parallel* to the magnetic field. However, if the electric and magnetic fields remain perpendicular to each other, this third term vanishes.

Now that we have a satisfactory solution for $g_k$, we can obtain the magnitude of the charge current by integrating the out-of-equilibrium population of electrons times velocity on the whole Fermi surface:

$$\mathbf{J}^e = \sigma E = \int v_k \, eg_k \, dk \qquad (4.20)$$

Assuming that the electric and magnetic field are perpendicular to each other, the expression for $g_k$ becomes

$$g_k = -\tau e \frac{\partial f^0}{\partial \epsilon_k} \mathbf{v_k} \cdot \frac{\mathbf{E} + \mu \mathbf{B} \times \mathbf{E}}{1 + \mu^2 B^2} \tag{4.21}$$

The last two equations yield

$$\mathbf{J^e} = e^2 \int \mathbf{v_k}(\tau) \left( -\frac{\partial f^0}{\partial \epsilon_k} \right) \mathbf{v_k} \cdot \frac{\mathbf{E} + \mu \mathbf{B} \times \mathbf{E}}{1 + \mu^2 B^2} dk \tag{4.22}$$

According to this equation, the current density vector $\mathbf{J^e}$ has now a component perpendicular to both electric field and magnetic field. Therefore, now it can be written as a sum of longitudinal, $\sigma_\|$, and transverse, $\sigma_\perp$, conductivities:

$$\mathbf{J^e} = \sigma_\| \mathbf{E} + \sigma_\perp (\mathbf{B}/\mathbf{B} \times \mathbf{E}) \tag{4.23}$$

The expression for the longitudinal conductivity is

$$\sigma_\| = e^2 \int (v_k^2 \tau) \left( -\frac{\partial f^0}{\partial \epsilon_k} \right) \frac{1}{1 + \mu^2 B^2} dk \tag{4.24}$$

While the transverse (i.e. Hall) conductivity can be written as

$$\sigma_\perp = e^2 \int (v_k^2 \tau) \left( -\frac{\partial f^0}{\partial \epsilon_k} \right) \frac{\mu B}{1 + \mu^2 B^2} dk \tag{4.25}$$

These expressions should be compared with Eq. 2.16 for zero-field conductivity, $\sigma_0$:

$$\sigma_0 = e^2 \int (v_k^2 \tau) \left( -\frac{\partial f^0}{\partial \epsilon_k} \right) dk \tag{4.26}$$

Assuming that mobility has no significant $k$-dependence, the two conductivities can be written as

$$\sigma_\| = \frac{\sigma_0}{1 + \mu^2 B^2} \tag{4.27}$$

$$\sigma_\perp = \frac{\sigma_0 \mu B}{1 + \mu^2 B^2} \tag{4.28}$$

We can see that the application of the magnetic field leads to a decrease in longitudinal conductivity. This is commonly called the magnetoresistance. It also induces an off-diagonal conductivity, which is called the Hall effect. Both quantities are set by mobility, a parameter, which we took as a scalar, but can be a tensor in general. Mobility is proportional to the scattering time and inversely proportional to the electronic effective mass. The lighter the quasi-particles and the longer their lifetime, the larger would be the field-induced change in the conductivity.

## 4.3   The Magneto-Seebeck and Nernst Effects

If instead of accompanying an electric field, the magnetic field is applied in the presence of a thermal gradient, the linearized Boltzmann equation will become (compare to Eq. 4.7)

$$\frac{\partial f_k^0}{\partial T} v_k \cdot \nabla T + \frac{g_k(r)}{\tau} + \frac{e}{\hbar}(v_k \times B) \cdot \nabla_k(g_k'(r)) = 0 \qquad (4.29)$$

The chain of arguments which led us from Eq. 4.6 to Eq. 4.10 can be used to extract from this equation the following expression for the distribution of electrons kicked out of equilibrium by a magnetic field:

$$g_k' = -\mathbf{v_k} \cdot \left[ \tau \left(\frac{\partial f^0}{\partial T}\right) \nabla \mathbf{T} + \mu \mathbf{B} \times \nabla_v(g_k') \right] \qquad (4.30)$$

Once again, the solution implies the search for a vector $\mathbf{G'}$, independent of velocity and defined as

$$\mathbf{G'} = \tau \left(\frac{\partial f_k^0}{\partial T}\right) \nabla \mathbf{T} + \mu \mathbf{B} \times \nabla_v(g_k') \qquad (4.31)$$

In the absence of a magnetic field, $\mathbf{G'}$ is $\mathbf{G_0'}$:

$$\mathbf{G_0'} = \tau \left(\frac{\partial f_k^0}{\partial T}\right) \nabla \mathbf{T} \qquad (4.32)$$

And in the presence of a magnetic field, $\mathbf{G'}$ should satisfy

$$\mathbf{G'} = \mathbf{G_0'} + \mu \mathbf{B} \times \mathbf{G'} \qquad (4.33)$$

This equation is identical to Eq. 4.13 and has exactly the same solutions. The expression for the out-of-equilibrium distribution of electrons in the presence of a thermal gradient and magnetic field is therefore straightforward:

$$g_k'(v) = -\tau \left(\frac{\partial f_k^0}{\partial T}\right) \mathbf{v_k} \cdot \frac{\nabla \mathbf{T} + \mu \mathbf{B} \times \nabla \mathbf{T} + (\mu^2 \mathbf{B} \cdot \nabla \mathbf{T})\mathbf{B}}{1 + \mu^2 B^2} \qquad (4.34)$$

We are now in a position to calculate thermoelectric response in the presence of a magnetic field. A finite charge current can be generated by a thermal gradient in the absence of an electric field. This happens if the thermoelectric response $\alpha$ is finite. The solution we found for $g'_k$ leads us to $\alpha$ using the relation:

$$\mathbf{J}^e = -\alpha \nabla T = \int v_k \, e g'_k \, dk \tag{4.35}$$

When the thermal gradient and magnetic field are perpendicular to each other, the third term of the right-hand side of Eq. 4.34 vanishes. Then the last two equations yield

$$\mathbf{J}^e = e \int \tau \mathbf{v_k} \left( \frac{\partial f^0_k}{\partial T} \right) \mathbf{v_k} \cdot \frac{\nabla T + \mu \mathbf{B} \times \nabla T}{1 + \mu^2 B^2} \, dk \tag{4.36}$$

We can separate the transverse and longitudinal components of $\alpha$:

$$\mathbf{J}^e = -\alpha_\parallel \nabla T - \alpha_\perp \left( \frac{\mathbf{B}}{B} \times \nabla T \right) \tag{4.37}$$

The longitudinal and transverse thermoelectric conductivities are thus

$$\alpha_\parallel = -e \int (\tau v_k^2) \left( \frac{\partial f^0}{\partial T} \right) \frac{1}{1 + \mu^2 B^2} \, dk \tag{4.38}$$

$$\alpha_\perp = -e \int (\tau v_k^2) \left( \frac{\partial f^0}{\partial T} \right) \frac{\mu B}{1 + \mu^2 B^2} \, dk \tag{4.39}$$

By assuming a featureless mobility, one is allowed to take it out of the integral. In that case, the longitudinal and transverse responses can be expressed in terms of the zero-field thermoelectric conductivity, $\alpha_0$:

$$\alpha_0 = e \int (\tau v_k^2) \frac{\partial f^0}{\partial T} \, dk \tag{4.40}$$

$$\alpha_\parallel = \frac{\alpha_0}{1 + \mu^2 B^2} \tag{4.41}$$

$$\alpha_\perp = \frac{\alpha_0 \mu B}{1 + \mu^2 B^2} \tag{4.42}$$

Thus, in the Boltzmann formalism, magnetic fields modify $\alpha$ much like $\sigma$. First of all, they modify the magnitude of the longitudinal thermoelectric response. This phenomenon is sometimes called the magneto-Seebeck effect. Moreover, a transverse thermoelectric response emerges and this is the origin of the Nernst effect.

## 4.4  Weak-Field and Strong-Field Limits

The Boltzmann equation in a magnetic field can be attacked by a method called Jones-Zener expansion [Jones and Zener 1934]. Consider the emergence of the following operator, $\Omega$, upon the application of the magnetic field:

$$\Omega = \overline{\mu}(\mathbf{v_k} \times \mathbf{B}) \cdot \nabla_v = \overline{\mu}\mathbf{v_k} \cdot (\mathbf{B} \times \nabla_v) \tag{4.43}$$

Using the definition of the operator $\Omega$, Eq. 4.9 can be written as

$$g_k(1 + \Omega) = \mathbf{v_k} \cdot \mathbf{E}\tau e \left( \frac{-\partial f_k^0}{\partial \epsilon_\mathbf{k}} \right) \tag{4.44}$$

Therefore, if one can find the inverse operator $(1+\Omega)$, the Boltzmann equation can be solved. Now, consider the series (see [Smith, Janak and Adler 1967])

$$(1 + \Omega)^{-1} = 1 - \Omega + \Omega^2 - \ldots \tag{4.45}$$

This is reminiscent of the series involving the real number $x$:

$$(1 + x)^{-1} = 1 - x + x^2 - \ldots \tag{4.46}$$

The latter converges only if $|x| < 1$. A similar situation occurs for the convergence of the series of operators in Eq. 4.45. If the magnetic field is too large, then the series cannot converge. The convergence criteria correspond roughly to $\mu B < 1$. At low enough fields, an approximate solution of the Boltzmann equation would be

$$g_k \approx (1 - \Omega + \Omega^2 - \ldots) \left[ -\mathbf{v_k} \cdot \mathbf{E}\tau e \left( \frac{-\partial f_k^0}{\partial \epsilon_\mathbf{k}} \right) \right] \tag{4.47}$$

As one can see from the definition of $\Omega$ (Eq. 4.43), what sets the order of magnitude of the change caused by this operator is the product of mobility and the magnetic field. Transport properties generally display distinct behaviour in the two extreme limits of magnetic field strength. When the magnetic field is a small perturbation to the electron trajectory ($\mu B << 1$), the transverse electric and thermoelectric conductivities are linear functions of a magnetic field and change sign as the magnetic field is inverted. The longitudinal conductivities, already finite at zero-field, display a small departure from this zero-field magnitude. They are indifferent to the orientation of a magnetic field and therefore are expected to vary proportional to $B^2$. These features are no more true at the other extreme of strong magnetic field ($\mu B >> 1$).

There is an alternative way of putting this. These two limits are distinct because they correspond to different ratios of scattering rate and cyclotron frequency. In the weak-field regime, the electron is scattered before completing one cyclotron period. In the strong-field regime, on the other hand, the cyclotron frequency is large and the cyclotron motion is not interrupted by scattering.

## 4.5   The Curvature of the Fermi Surface and the Sign of the Transport Coefficients

In the last section of Chapter 2, we saw that the sign of the Seebeck coefficient is linked to the curvature of the Fermi surface or the sign of carriers. A Fermi surface of free electrons, in other words, a filled sphere, has a negative Seebeck coefficient. This is because the overall number of occupied states just below the Fermi level is lower than the overall number of the unoccupied states just above the Fermi level. In the case of a shallow Fermi surface, the sign is opposite because the filled states outnumber the empty states.

The curvature of the Fermi surface determines the sign of the Hall coefficient as well. However, the mechanism leading to opposite signs for hole-like and electron-like Fermi surfaces is not the same. As sketched in Figure 4.4, in the electron-like case, the out-of-equilibrium depopulation vector of electrons is skewed in an orientation opposite to the hole-like case. The sign of the Hall response depends on the local curvature of the Fermi surface. If it bends inward, the transverse electric field will be opposite to the case it bends outward.

Thus, holeness is subtly different seen from the Seebeck and the Hall point of views. In the case of the Seebeck response, what matters is the balance of population between occupied and unoccupied states. In the case of the Hall response, the sign is set by the sign (inwards or outwards) of the velocity (see [Ong 1991] for an elegant and rigorous formulation of the link between Hall conductivity and Fermi surface curvature). Such a difference does not matter in the two simple cases of full or shallow spheres.

In the absence of anomalous scattering, the local curvature of the Fermi surface is expected to lead to identical signs for the Hall and Seebeck coefficients in the case of filled and shallow spheres. The situation is much less clear when the Fermi surface has a less

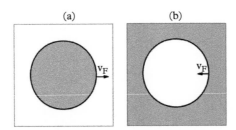

**Figure 4.4** *The opposite signs of the Hall response for electrons and holes. (a) In an electron-like surface, the Fermi velocity is outwards. (b) In a hole-like Fermi surface, the Fermi velocity is inwards. Therefore, the transverse electric fields generated by the application of a magnetic field have opposite signs.*

trivial shape. Moreover, the sign of the Seebeck coefficient can be altered if the mean-free-path has a strong downward energy dependence. If the difference in the scattering rate between empty and filled states outweighs the difference between their populations, the sign of the Seebeck coefficient could be inverted. We will come back to this while discussing the sign of the Seebeck coefficient in noble metals.

What about the sign of the Nernst effect? As sketched in Fig. 4.4, the Nernst response, like the Seebeck case, is set by the difference between contributions of occupied and un-occupied states. On the other hand, like the Hall case, the electrons travelling opposite the thermal gradient contribute with an opposite sign to those travelling along the ther-mal gradient. Therefore, the sign of the Nernst response of the two (shallow and filled) Fermi surfaces is identical, contrary to both Seebeck and Hall coefficients. See Fig. 4.5 for a representation of the relevant contributions to the four transport coefficients.

Many real metals have complex Fermi surfaces with several components, each with both positive and negative local curvatures. The sign of the Hall coefficient is set by a subtle sum of these opposite contributions. The same is true for the Seebeck effect. The

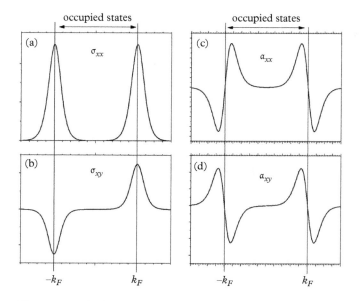

**Figure 4.5** *The four transport coefficients probe different electrons.*
*(a) Longitudinal conductivity is set by electrons at the Fermi surface.*
*(b) Hall conductivity is set by a difference between the contribution of leftward and rightward travelling electrons at the opposite ends of the Fermi surface. (c) Longitudinal thermoelectric response is set by the difference between the contribution of occupied and unoccupied states. (d) Transverse thermoelectric response is set by the difference between the contribution of occupied and unoccupied states, but with a different sign at the opposite ends of the Fermi surface.*

curvature does not alter these two physical properties in the same way. In many multi-band metals, the signs of the Hall and Seebeck coefficients do not match and they can present a Nernst coefficient of either sign.

As we will see in our experimental survey, even when the topology of the Fermi surface is simple, a detailed knowledge of this topology does not suffice to understand the sign of the Seebeck coefficient, and one is tempted to invoke a complex energy dependence for scattering time. In the case of Hall coefficient, on the other hand, this is not the case, at least in the case of most elemental metals [Ziman 1960].

# 5

# The Thermal Wave-Length and Fermi-Liquid Thermoelectricity

The aim of this chapter is to draw a simple picture of thermoelectric response in a Fermi liquid. The extreme simplicity has the merit of highlighting the ultimate source of entropy carried by mobile electrons. In a Fermi-Dirac distribution, entropy is scarce and is narrowly located near the singularity of the distribution. This has significant implications for the magnitude of the Seebeck coefficient. The latter is set in the final instance by the de Broglie thermal wavelength of electrons.

## 5.1 The Fermi Liquid

Nowadays, the standard theory of metals is often dubbed 'The Fermi liquid picture', in reference to a concept devised by Lev Landau. The band theory successfully predicts the ground state of most solids, in spite of wrongly assuming that electrons are invisible to each other. The concept of Fermi liquid provides a basis to understand this amazing fact. A Landau Fermi liquid is composed of quasi-particles, which are not but behave like free electrons. The neglected Coulomb repulsion among electrons is translated to the normalized mass of quasi-particles. As a consequence, the non-interacting theory can punch above its weight. It remains correct in its broad lines, provided that one assumes that interactions would only enhance the quasi-particle mass.

This is not always the case. Systems in which the Fermi-liquid picture is suspected to fail are a recurrent focus of contemporary investigation. Experiments are routinely performed on these strange metals, which emerge in the vicinity of a quantum critical point or when a Mott insulator is doped to extract properties falling out of the scope of the Fermi-liquid picture.

In this chapter, we focus on the transport response of isotropic Fermi liquids in the zero-temperature limit. In Chapter 2, we reviewed the semi-classical picture of electron transport. We saw that the transport properties of a given metal are set by the structure of its Fermi surface and the (eventually anisotropic) scattering time of electrons. In this chapter, we consider the idealized case of a Fermi liquid with a spherical (or a cylindrical) Fermi surface. Its unique difference, with a free-electron gas, is the effective mass of its building entities, which could be different from the bare electron mass.

*Fundamentals of Thermoelectricity*. First Edition. Kamran Behnia.

This effective mass quantifies the way rising temperature smears the Fermi surface in question. We will see that, in the low-temperature limit, charge and thermal conductivities can be expressed in terms of three fundamental constants (the Planck constant $\hbar$, the electron charge $e$, and the Boltzmann constant $k_B$) and three material-dependent length scales (the Fermi wave-vector $\lambda_F$, the de Broglie thermal wave-vector $\lambda_{dB}$, and the electron mean-free path $\ell$).

## 5.2   The Quantization of Conductance

The natural unit of electric conductance is $G_0 = \frac{e^2}{h} \simeq 3.87 \times 10^{-5}$ S. This fact has become widely acknowledged following two parallel scientific evolutions in condensed-matter physics during the last two decades of the twentieth century. The first is the discovery of the quantum Hall effect. The Hall conductance of a two-dimensional electron gas becomes quantized in integers of $G_0$. The second is the elaboration of the modern quantum transport theory by Landauer, which laid the foundation of understanding conduction as transmission [Imry and Landauer 1999].

Such a natural unit does not appear in the semi-classic theory of electron transport. Consider the familiar expression for charge conductivity in this theory:

$$\sigma = \frac{ne^2\tau}{m^*} = ne\mu \tag{5.1}$$

Let us have a look at the components of this equation. Conductivity is proportional to the carrier concentration $n$, which implies a better conduction with more carriers. It is proportional to $e$, because this is the amount of electric charge a carrier transports. Finally, the mobility $\mu$ quantifies the capacity of an average carrier to transport. The mobility is proportional to the product of two length scales, the mean-free-path and the electron wavelength. The latter is inversely proportional to momentum. Thus, a carrier is more mobile if it can carry its charge over a longer distance with a lower momentum.

All this makes sense, but this equation does not define any limits for the conductivity of a metal. It can take any value from zero to infinity. Now, consider the following reformulation of Eq. 5.1 in a three-dimensional metal with a spherical Fermi radius of $k_F$:

$$\sigma = \frac{2}{3\pi} \frac{e^2}{h} k_F^2 \ell \tag{5.2}$$

What we have done is just replace the carrier density and mobility by their equivalent expressions. The emergence of the quantum of conductance allows us to see the natural unit. Charge conduction is about transmitting units of $G_0$, the two material-related properties of the radius of the Fermi sphere, $k_F$, and the mean-free-path, $\ell$. Recall that all electrons do not participate in charge transport, but only those which are at the Fermi surface. In three dimensions, the number of such electrons scales with the square of the Fermi radius; hence the exponent.

**Table 5.1** *Charge conductivity in different dimensions.*

| Dimensions | 3D | 2D | 1D |
|---|---|---|---|
| | $\dfrac{2}{3\pi}\dfrac{e^2}{h}k_F^2\ell$ | $\dfrac{e^2}{h}k_F\ell$ | $4\dfrac{e^2}{h}\ell$ |
| Units | $\mathrm{S\,m^{-1}}$ | S | $\mathrm{S\,m}$ |

Table 5.1 summarizes the expressions for conductivity for one, two, and three dimensions. For each dimension $D$, the transmission probability is proportional to the mean-free-path and a dimensional-dependent coefficient proportional to $k_F^{D-1}$.

Less well-known than $G_0$ are other combinations of the fundamental constants giving rise to natural units for thermal and thermoelectric conductivities. They are listed in Table 2.1. The natural unit for thermoelectric conductance, $\alpha$, is $\frac{ek_B}{h}$. For thermal conductance divided by temperature, $\kappa/T$, it is $\frac{k_B^2}{h}$. It follows that the natural unit for the Seebeck coefficient, the ratio of $\alpha$ to $\sigma$, is $\frac{k_B}{e}$, the ratio of natural units of $\alpha$ and $\sigma$. In Chapter 2, we saw how these natural units emerge in Landauer formalism. In this chapter, we are going to find expressions for $\alpha$, $\kappa/T$, and $S$ expressed in their natural units. These would be thermoelectric and thermal equivalents to Eq. 5.2. But, before this, we should add a new length scale to the other two.

## 5.3 Three Length Scales

In the previous section, we saw that by expressing electrical conductivity in its natural units, one can see that there are two material-dependent length scales: the Fermi wavelength of electrons $\lambda_F$ and the mean-free-path $\ell$. Following this thread, let us introduce a third length scale, the thermal de Broglie wavelength, defined as

$$\lambda_{dB} = \sqrt{\frac{2\pi\hbar^2}{m^* k_B T}} \tag{5.3}$$

Unlike the Fermi wavelength, $\lambda_{dB}$ varies with temperature. When the temperature is low enough for it to become longer than the average distance between fermions, the Fermi gas becomes degenerate. Thus, at the Fermi temperature, the thermal wavelength $\lambda_{dB}$ of each particle becomes equal to the average inter-particle distance. This is one definition of this length scale. The Heisenberg uncertainty principle provides another one. The thermal wavelength is a measure of the thermodynamic uncertainty in the localization of a particle whose momentum is set by thermal energy [Silvera 1997]. Note that the heavier the quasi-particles, the shorter their thermal de Broglie wavelength. Note also that $\lambda_{dB}$ diverges, at zero temperature.

It is useful to define the counterparts of these three length scales in the reciprocal space. The Fermi wavelength is inversely proportional to the Fermi wave-vector.

The latter is the radius of the Fermi sphere in the reciprocal space. One can also define a thermal de Broglie wave-vector thus:

$$k_{dB} = \frac{2\pi}{\lambda_{dB}} \tag{5.4}$$

What is the physical meaning of this quantity in the reciprocal space? It vanishes at absolute zero. At a high enough temperature, it becomes larger than the Fermi wave-vector. Interestingly, when the system is warmed to its Fermi temperature, the two wave-vectors become equal. Above this temperature, the Fermi liquid becomes non-degenerate. In the degenerate regime, $k_{dB}$ is smaller than $k_F$ and quantifies the thermal thickness of the Fermi surface. The larger the $k_{dB}$, the fuzzier is the Fermi sphere and the broader is the transition between the occupied and unoccupied states.

The mean-free-path has also a counterpart in the reciprocal space. If the electron wave is restricted to a space the size of $\ell_e$ then, according to the Heisenberg uncertainty principle, its momentum $\hbar k$ is well defined within $\frac{h}{\ell}$. Let us define the reciprocal wavelength $k_s$:

$$k_s = \frac{2\pi}{\ell} \tag{5.5}$$

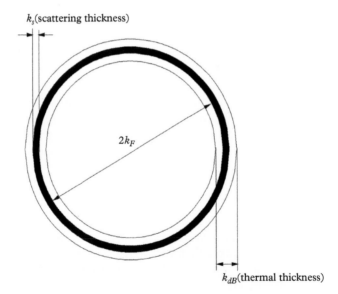

Figure 5.1 *The three length scales in the reciprocal space. The transport properties can be expressed in terms of the quanta of conductance and the three length scales. The variable $k_F$ is the radius of the Fermi sphere, $k_{dB}$ is its thermal thickness, set by temperature and the effective mass, and $k_s$ is the scattering thickness of the Fermi surface set by the mean-free-path of electrons.*

It quantifies how finely the $k$-states of the reciprocal space are defined. At the frontier between occupied and unoccupied states, there is a 'scattering thickness' of the Fermi surface represented by the finite value of $k_s$. Figure 5.1 schematically sketches these three characteristic scales in the reciprocal space.

If we replace the mean-free-path by scattering thickness in Eq. 5.2, it becomes

$$\sigma = \frac{4}{3}\frac{e^2}{h}\frac{k_F^2}{k_s} \tag{5.6}$$

The link to Landauer's original idea of conductance as transmission is now transparent. Charge conductivity is set by the quantum of conductance. It is proportional to $k_F^2$, because the size of the outer surface quantifies the number of excitable states and is inversely proportional to the scattering thickness, because of the number of effective excitable states inversely scales with the minimal size of a well-defined state.

All this happens in the reciprocal space.

## 5.4   Thermoelectric and Thermal Conductivities

Consider Eqs 2.41–43, which contain the transport distribution function [Mahan and Sofo 1996], which can be designated by $\Xi(\epsilon)$. Assuming a constant mean-free-path and a spherical Fermi surface, Eq. 2.36 yields

$$\Xi(\epsilon) = k^2 l/3\pi = \frac{2m^*\epsilon}{3\pi\hbar^2}\ell \tag{5.7}$$

To extract an expression for thermoelectric conductivity, we need to find the energy derivative of this function at the Fermi energy:

$$k_B T\frac{\partial \Xi}{\partial \epsilon}\Big|_{\epsilon=\epsilon_F} = \frac{2m^*k_B T}{3\pi\hbar^2}\ell = \frac{4}{3}\frac{\ell}{\lambda_{dB}^2} \tag{5.8}$$

Using Eq. 2.42, this leads to

$$\alpha = \frac{8\pi^2}{9}\frac{ek_B}{h}\frac{\ell}{\lambda_{dB}^2} \tag{5.9}$$

or alternatively,

$$\alpha = \frac{2}{9}\frac{ek_B}{h}k_{dB}^2\ell \tag{5.10}$$

As for thermal conductivity, one can use Eqs 2.43 and 5.7 to find

$$\kappa/T = \frac{2\pi}{9}\frac{k_B^2}{h}k_F^2\ell \tag{5.11}$$

Replacing the mean-free-path by its $k$-space counterpart $k_s$, one can rewrite the two last equations as

$$\alpha = \frac{4\pi}{9} \frac{ek_B}{h} \frac{k_{dB}^2}{k_s} \tag{5.12}$$

$$\kappa/T = \frac{4\pi^2}{9} \frac{k_B^2}{h} \frac{k_F^2}{k_s} \tag{5.13}$$

Thermoelectric conductivity is not set by the radius of the Fermi sphere but by its thermal fuzziness. The latter quantifies the number of states available for exploration to electrons in the vicinity of the Fermi surface. On the other hand, thermal conductivity is set by the radius of the Fermi sphere, very much like electrical conductivity. The scattering thickness is relevant because it quantifies the number of available states.

## 5.5 Dimensionality

The number of $k$-states that one can put inside an energy window depends on the dimensionality of the system. Therefore, the energy dependence of the transport distribution function is not the same in two and three dimensions. Replacing the spherical Fermi surface of radius $k_F$ with a cylindrical Fermi surface of radius $k_F$ and height $a_k$, we find a new expression for the energy dependence of $\Xi$ (see Fig. 5.2):

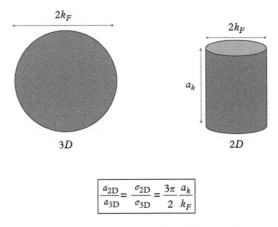

$$\frac{a_{2D}}{a_{3D}} = \frac{\sigma_{2D}}{\sigma_{3D}} = \frac{3\pi}{2} \frac{a_k}{k_F}$$

**Figure 5.2** *Spherical vs cylindrical Fermi surfaces. Transport properties depend on dimensionality. Both electric and thermoelectric conductivities are enhanced in the case of a cylindrical Fermi surface with a height $a_k$ equal to the reciprocal lattice length.*

**Table 5.2** *Transport transmission coefficients in 2D and 3D.*

|  | 2D | 3D |
|---|---|---|
| $\sigma$ | $k_F \ell$ | $k_F^2 \ell$ |
| $\alpha$ | $(k_{dB}^2/k_F)\ell$ | $k_{dB}^2 \ell$ |
| $\kappa/T$ | $k_F \ell$ | $k_F^2 \ell$ |

$$\Xi^{2D}(\epsilon) = \frac{a_k k}{2}\ell = \frac{a_k \sqrt{2m^*\epsilon}}{2\hbar}\ell \tag{5.14}$$

This leads us to the following expressions for, electric, thermoelectric, and thermal conductivities:

$$\sigma^{2D} = \frac{e^2}{h} a_k k_F \ell \tag{5.15}$$

$$\alpha^{2D} = \frac{\pi}{3}\frac{ek_B}{h}\frac{a_k}{k_F}k_{dB}^2 \ell \tag{5.16}$$

$$\kappa^{2D}/T = \frac{\pi^2}{3}\frac{k_B^2}{h} a_k k_F \ell \tag{5.17}$$

Note that these expressions depend on the unit length of the reciprocal lattice $a_k$. It is instructive to compare the magnitude of thermoelectric conductivities in $2D$ and $3D$:

$$\frac{\alpha^{2D}}{\alpha^{3D}} = \frac{3\pi}{2}\frac{a_k}{k_F} \tag{5.18}$$

Since $a_k$ is larger than $k_F$, this implies that such a decrease in dimensionality enhances the thermoelectric conductivity. As we will see below, it would not affect the Seebeck coefficient.

Table 5.2 compares transmission coefficients in two and three dimensions. It is striking that $\alpha$ is the single transport coefficient, which depends on $k_{dB}$. Note also that the exponent of $k_F$ depends on the dimensionality of the system, but the exponent of $k_{dB}$ does not. Available entropy scales with $k_{dB}^2$ independent of entropy.

## 5.6   The Scattering-Independent Seebeck Coefficient

The expression for electric and thermal conductivities obtained in the previous section points to a fundamental ratio:

$$\frac{\sigma}{\kappa/T} = \frac{\pi^2}{3} \left( \frac{k_B}{e} \right)^2 \tag{5.19}$$

This is, of course, the Wiedemann-Franz law. The ratio of heat-to-charge conductivity is set by the ratio of Boltzmann constant to the charge of the electron. In other words, while thermal and electric conductivity both depend on the mean-free-path, their ratio does not. It is a universal constant. Note also that ratio holds in both cases of the spherical and the cylindrical Fermi surface.

Now, while both the electric and thermoelectric conductivities are proportional to the mean-free-path, their ratio is not. Combining Eqs 5.2 and 5.10 yields the Seebeck coefficient:

$$S = \frac{\alpha}{\sigma} = \frac{\pi}{3} \frac{k_B}{e} \left( \frac{k_{dB}}{k_F} \right)^2 \tag{5.20}$$

Within this approximation (a mean-free-path with no energy dependence), in a Fermi liquid cooled well below its Fermi temperature, the Seebeck coefficient does not depend on scattering. It is set by the ratio of two material-dependent wave-vectors. It is also noteworthy that combining equations relevant to the $2D$ case (Eqs 5.15 and 5.16) leads to the same result. In other words, dimensionality, which alters both $\sigma$ and $\alpha$, does not affect the expression of the Seebeck coefficient.

This brings us back to the paradox of the Seebeck coefficient. It is a transport property, but its magnitude becomes independent of how far the electrons can travel. This feature distinguishes thermopower from other transport properties of a Fermi liquid.

Eq. 5.20 is strictly equivalent to the following more familiar one:

$$S = \frac{\pi^2}{3} \frac{k_B}{e} \frac{T}{T_F} \tag{5.21}$$

In a simple Fermi liquid, the Seebeck coefficient is expected to be linear in temperature and the slope of this $T$-linear variation inversely scales with the magnitude of the Fermi temperature.

As we will see in the next chapters, this fundamental property of a Fermi liquid is hard to check experimentally, at least in simple metals [Macdonald 1962; see Chapter 6]. Ironically, it is in strongly correlated metals in general and in heavy fermions in particular that this simple equation has a striking experimental relevance [Behnia, Jaccard, and Flouquet 2004; see Chapter 8]. On the other hand, its experimental verification with high accuracy has been reported in two metallic single-band semiconductors, namely heavily doped silicon [Lakner and Löhneysen 1993; see Chapter 6] and in metallic $SrTiO_{3-\delta}$ [Lin et al. 2013; see Chapter 8]. Both are single-band metals with a tiny Fermi surface.

## 5.7    Energy-Dependent Scattering

In deriving the expression for thermoelectric conductivity, $\alpha$ (Eqs 5.9–10), we assumed that the mean-free-path is featureless and its energy derivative is therefore zero.

This need not be the case in general. Taking into consideration such a possibility, we can rewrite the energy derivative of the transport distribution function as

$$k_B T \frac{\partial \Xi}{\partial \epsilon} \Big|_{\epsilon=\epsilon_F} = \frac{8\pi m^* k_B T}{3h^2} \ell \left[ 1 + \frac{\partial ln(\ell)}{\partial ln(\epsilon)} \Big|_{\epsilon=\epsilon_F} \right] \tag{5.22}$$

In this case, Eq. 5.10 would lose its simple elegance, becoming

$$\alpha = \frac{8\pi^2}{9} \frac{e k_B}{h} k_{dB}^2 \ell \left[ 1 + \frac{\partial ln(\ell)}{\partial ln(\epsilon)} \Big|_{\epsilon=\epsilon_F} \right] \tag{5.23}$$

The Seebeck coefficient, in turn, would include an additional term quantifying the importance of the energy dependence of the mean-free-path:

$$S = \frac{\pi}{3} \frac{k_B}{e} \left( \frac{k_{dB}}{k_F} \right)^2 \left[ 1 + \frac{\partial ln(\ell)}{\partial ln(\epsilon)} \Big|_{\epsilon=\epsilon_F} \right] \tag{5.24}$$

Consequently, Eq. 5.21 should be revised to

$$S = \frac{\pi^2}{3} \frac{k_B}{e} \frac{T}{T_F} \left[ 1 + \frac{\partial ln(\ell)}{\partial ln(\epsilon)} \Big|_{\epsilon=\epsilon_F} \right] \tag{5.25}$$

One can directly deduce this equation from the Mott equation by considering that the energy dependence of charge conductivity has two distinct components, only one of which (here, the first term of the right side) is purely thermodynamic.

Even with a simple single-band Fermi surface with no anisotropy, the scattering term may significantly modify the Seebeck coefficient. Let us suppose that the mean-free-path is a smooth function of energy. In that case, near the Fermi energy, it can be written as

$$\ell(\epsilon) = \ell_0 \epsilon^\lambda \tag{5.26}$$

The exponent is related to the scattering term

$$\lambda = \frac{\partial ln(\ell)}{\partial ln(\epsilon)} \Big|_{\epsilon=\epsilon_F} \tag{5.27}$$

The simplest case corresponds obviously to $\lambda = 0$ (i.e. a constant mean-free-path). In this case, all electrons travel the same distance before being scattered. One may suppose this if scattering centres are hard-sphere impurities, scattering electrons independent of their energy. A second simple case is when the scattering rate does not depend on the energy of electrons. Each scattering event occurs within a constant time window. This is often called the constant relaxation-time approximation. The mean-free-path $\ell$ and the scattering time $\tau$ are linked through the Fermi velocity:

$$\ell(\epsilon) = \tau(\epsilon) v_\epsilon \tag{5.28}$$

In this approximation, each electron travels for the same temporal interval before being scattered. Hence the more energetic electrons, which are faster, travel longer than the slower less energetic electrons. It is the energy dependence of velocity, which sets $\lambda$:

$$\lambda = \frac{\partial ln(v(\epsilon))}{\partial ln(\epsilon)} = 1/2 \tag{5.29}$$

In that case, the Seebeck coefficient would become

$$S = \frac{\pi^2}{2} \frac{k_B}{e} \frac{T}{T_F} \tag{5.30}$$

In addition to these simplest cases, one can imagine many other scenarios. If the scattering is such that the more energetic electrons are scattered harder and travel a shorter distance, then $\lambda$ can be negative. In the case that this feature becomes so pronounced that $\lambda < 1$, even the sign of the Seebeck coefficient can be affected. As we will see in the next chapter, this is what has been suggested to explain the positive Seebeck coefficient of noble metals.

However, in almost all realistic scenarios, that is, in all cases where the mean-free-path is a smooth function of energy, $\lambda$ is expected to wander not far from unity. Basically, Eq. 5.26 implies that $\ell \propto \epsilon^\lambda$. Therefore, the following simple expression for the Seebeck coefficient is quite general:

$$S \simeq A \frac{T}{T_F} \tag{5.31}$$

Since in most known cases $\lambda$ is of the order of unity, the prefactor $A = \frac{\pi^2}{3}(1 + \lambda)\frac{k_B}{e}$ is expected to wander around $\frac{\pi^2}{3}\frac{k_B}{e} = 283\ \mu VK^{-2}$.

A noteworthy exception is the single-particle Kondo effect, in which $\lambda$ is temperature dependent. This temperature dependence is strong enough to overthrow the usual $T$-linear behaviour.

## 5.8   Seebeck Coefficient and Specific Heat

The Fermi energy of a gas of Fermions sets the magnitude of its specific heat. In thermodynamics, one defines the heat capacity of a solid body as the amount of heat needed to increment its temperature. The larger the body, more thermal energy is needed to enhance its temperature by a unit of temperature (often a Kelvin degree). The specific heat of a given material is its heat capacity by unit of mass or volume. The contribution of electrons to the specific heat of a solid body is [Ziman 1960]

$$C_{el} = \frac{\pi^2}{3} k_B^2 T N(\epsilon_F) \tag{5.32}$$

Here, the function $N(\epsilon)$ quantifies the density of states at the energy $\epsilon$. In $k$-space, the number of available states per unit $k$ is constant. The density of states at the Fermi level can be shown to be [Ziman 1960]

$$N(\epsilon_F) = \frac{A_F}{4\pi^3 \hbar v_F} \qquad (5.33)$$

In plain words, the density of states is set by the ratio of the Fermi area $A_F$ to the Fermi velocity $v_F$. In the case of a Fermi sphere, the former is

$$A_F = 4\pi k_F^2 \qquad (5.34)$$

and the Fermi velocity can be expressed as the ratio of the Fermi wave-vector to the effective mass:

$$v_F = \frac{\hbar k_F}{m^*} \qquad (5.35)$$

Using Eqs 5.33–35, Eq. 5.32 can be written as

$$C_{el}/T = \frac{1}{3} \frac{k_B^2}{\hbar^2} k_F m^* \qquad (5.36)$$

Using Eq. 5.3, this equation can be rewritten as

$$C_{el} = \frac{\pi^2}{2} k_B \frac{k_F}{k_{dB}^2} \qquad (5.37)$$

An alternative expression can be obtained by noting that, in a free-electron gas, the density of states is set by volume density of carriers $n$ and the Fermi energy:

$$N(\epsilon_F) = \frac{2}{3} \frac{n}{\epsilon_F} \qquad (5.38)$$

Injecting this into Eq. 5.32 leads to a more familiar expression for specific heat:

$$C_{el} = \frac{\pi^2}{2} k_B \frac{nT}{T_F} \qquad (5.39)$$

Thus, the electronic specific heat is a measure of the ratio of carrier concentration to Fermi energy. This equation is strikingly similar to Eqs 5.21 and 5.30. While the Seebeck coefficient measures entropy per carrier, the specific heat measures entropy per volume. In the simplest case, their ratio is set by carrier density. Specific heat is usually expressed per one mole of the material and its standard unit is $JK^{-1} mol^{-1}$. Let us introduce a dimensionless ratio of the Seebeck coefficient to the electronic specific heat [Behnia, Jaccard, and Flouquet 2004]:

$$q = \frac{SN_{av}e}{C_{el}} \tag{5.40}$$

where $N_{av}$ is the Avogadro number and $e$ is the charge of the electron. For a free-electron gas with a constant relaxation time (Eq. 5.30), one can expect $q = -1$. In the case of a constant mean-free-path (Eq. 5.21), one expects $q = 2/3$. More generally, in the case of a one-band system with an energy-dependent mean-free-path, combining Eqs 5.25 and 5.39, one has

$$q = \frac{2}{3}\frac{1}{n}(1 + \lambda) \tag{5.41}$$

The parameter $\lambda$ quantifies the energy dependence of the mean-free-path. Its order of magnitude cannot wander much far from unity. Thus, for any solid with roughly one electron per formula unit, $q$ is expected to be of the order of unity. However, if $\lambda$ becomes negative and its magnitude exceeds unity, the sign of the Seebeck coefficient becomes opposite to what is expected. We will come back to this possibility when we discuss the sign of the Seebeck coefficient in noble metals.

## 5.9    Magnetic Length and Off-Diagonal Coefficients

In the presence of a magnetic field, the three conductivities $\bar{\sigma}$, $\bar{\kappa}$, and $\bar{\alpha}$ are tensors, which relate charge current, $\mathbf{J}^e$, and heat current, $\mathbf{J}^Q$ to electric field, $\mathbf{E}$, and thermal gradient, $\nabla \mathbf{T}$:

$$\mathbf{J}^e = \bar{\sigma} \cdot \mathbf{E} - \bar{\alpha} \cdot \nabla \mathbf{T} \tag{5.42}$$

$$\mathbf{J}^Q = T\bar{\alpha} \cdot \mathbf{E} - \bar{\kappa} \cdot \nabla \mathbf{T} \tag{5.43}$$

Let us turn our attention to $\sigma_{xy}$ and $\alpha_{xy}$. What sets their magnitude? In order to answer this question we should introduce a new length scale, commonly called the magnetic length:

$$\ell_B^2 = \frac{\hbar}{eB} \tag{5.44}$$

We discussed the consequences of the application of a magnetic field for electric and thermoelectric transport in Chapter 3. The crucial parameter was found to be the electron mobility, $\mu$. In the weak-field regime, defined by $\mu B \ll 1$, one has $\sigma_\perp = \mu B \sigma_\parallel$. Therefore, the linear growth of the off-diagonal terms with a magnetic field depends on the mobility of the system. However, the product $\mu B$ is also a ratio of length scales:

**Table 5.3** *Natural units and characteristic length for off-diagonal transport coefficients in 2D and 3D.*

|  | Natural units | 2D length | 3D length |
|---|---|---|---|
| $\sigma_{xy}/B$ | $\frac{e^3}{h^2}$ | $\ell^2$ | $k_F \ell^2$ |
| $\alpha_{xy}/B$ | $\frac{k_B e^2}{h^2}$ | $\left(\frac{k_{dB}}{k_F}\right)^2 \ell^2$ | $\frac{k_{dB}^2}{k_F} \ell^2$ |
| $\kappa_{xy}/BT$ | $\frac{k_B^2 e}{h^2}$ | $\ell^2$ | $k_F \ell^2$ |

$$\mu B = \frac{\ell}{k_F \ell_B^2} \tag{5.45}$$

Using this and Eq. 5.2, one can write down an expression for the Hall conductivity of a spherical Fermi surface:

$$\sigma_{xy} = \frac{2}{3\pi} \frac{e^2}{h} \frac{k_F \ell^2}{\ell_B^2} \tag{5.46}$$

A similar expression can be written for the off-diagonal Peltier conductivity:

$$\alpha_{xy} = \frac{2}{9} \frac{e k_B}{h} \frac{k_{dB}^2 \ell^2}{k_F \ell_B^2} \tag{5.47}$$

In the low-field regime, off-diagonal coefficients would be linear in a magnetic field. Therefore, their natural units would be the quanta of conductance divided by the quantum of the magnetic flux, the ratio $\frac{e}{h}$. The transmission coefficient is a square of length in 2D and a simple length in 3D. Table 5.3 summarizes these units. It can be seen that, in the case of $\alpha_{xy}$, the magnitude is set by a combination of mean-free-path, Fermi wavelength, and thermal wavelength.

## 5.10 The Nernst Coefficient

A rough estimate of the magnitude of the Nernst coefficient in a Fermi liquid can be useful for understanding the experimental data at very low temperatures [Behnia 2009]. In the absence of a charge current (i.e. $\mathbf{J^e} = 0$):

$$\mathbf{E} = \overline{\sigma}^{-1} \cdot \overline{\alpha} \cdot \nabla \mathbf{T} \tag{5.48}$$

The Nernst signal, the transverse electric field $E_y$, generated by a longitudinal thermal gradient $\nabla_x T$, is

$$N = \frac{E_y}{\nabla_x T} = \frac{\alpha_{xy}\sigma_{xx} - \alpha_{xx}\sigma_{xy}}{\sigma_{xx}^2 + \sigma_{xy}^2} \tag{5.49}$$

Using the definition of the Hall angle $(\tan\theta_H = \frac{\sigma_{xy}}{\sigma_{xx}})$, one can write this as [Oganesyan 2004; Behnia 2009]

$$N = -\frac{\pi^2}{3}\frac{k_B^2 T}{e}\frac{\partial \tan\theta_H}{\partial\epsilon}\Big|_{\epsilon=\epsilon_F} \tag{5.50}$$

Thus, the Nernst coefficient is linked to the energy derivative of the Hall angle, $\frac{\partial \tan\theta_H}{\partial\epsilon}\big|_{\epsilon=\epsilon_F}$. In most cases, the Hall angle is a smooth function of energy in the vicinity of the chemical potential. In these cases, the energy derivative is just $\frac{\tan\theta_H}{\epsilon_F}$ times a numerical factor close to unity; this is equivalent to assuming that $\tan\theta_H$ is a linear function of energy. In a one-band metal, the Hall angle measures the carrier mobility, $\mu$. The latter can be expressed as

$$\tan\theta_H/B = \mu = \frac{e\tau}{m^*} = \frac{e\ell}{\hbar k_F} \tag{5.51}$$

Here, $k_F$ is the Fermi wave-vector, and $\ell$ is the carrier mean-free-path. This is particularly useful in the case of multi-band metals. Since the sign of the Hall angle is different for hole-like and electron-like carriers, the overall Hall angle of an ambipolar metal can be substantially reduced compared to the Hall angle (and the mobility) of each band.

These two simplifications lead us to the following expression for the magnitude of the Nernst coefficient:

$$\nu = \frac{\pi^2}{3}\frac{k_B}{e}\frac{k_B T}{\epsilon_F}\mu \tag{5.52}$$

## 5.11   Multiple Bands and Multiple Scattering Mechanisms

What if there are several bands? What if there are several different ways by which electrons can lose their momentum and/or energy? All the discussion in the preceding sections of this chapter was focused on a one-band system with featureless scattering. Let us consider briefly the modifications introduced as soon as the situation becomes more complex.

In a multi-band system, one would expect the contributions by each band to sum up:

$$\sigma = \sum_i \sigma_i \tag{5.53}$$

$$\alpha = \sum_i \alpha_i \tag{5.54}$$

Consider a two-band system. Band $a(b)$ has a single-band Seebeck coefficient of $S_a(S_b)$ and an electric conductivity of $\sigma_a(\sigma_b)$. Eq. 5.54 implies the following expression for the total Seebeck coefficient of the system:

$$S = \frac{S_a \sigma_a + S_b \sigma_b}{\sigma_a + \sigma_b} \qquad (5.55)$$

There is no particular assumption behind this equation regarding the microscopic scattering processes. It is a consequence of the fact that $\sigma$ and $\alpha$ are the building blocks of the fundamental transport equations.

Now, let us consider an alternative case story. What happens if, in a single-band system, there are several distinct scattering mechanisms that we can identify? In this case, the resistivity can be written as

$$\rho = \sum_i \rho_i \qquad (5.56)$$

Here, each index $i$ refers to a particular scattering mechanism. Each electron has a finite probability to be scattered by an impurity, by a phonon, or by another electrons. These probabilities simply add up. Thus, the scattering rate and, as a consequence, the total resistivity is a sum of contributions by these different scattering mechanisms.

The Nordheim-Gorter [Nordheim and Gorter 1935] rule states that, in such a context, the total Seebeck coefficient of the system is

$$S = \sum_i \frac{\rho_i S_i}{\sum_i \rho_i} \qquad (5.57)$$

Such an equation can be deduced in two different ways. It can be obtained by considering successive scattering events of different types as a serial circuit connected together with a specific temperature gradient, $\Delta T_i$, each. In that case,

$$\Delta T_i = \frac{W_i \Delta T}{\sum_i W_i} \qquad (5.58)$$

where $W_i$ is the thermal resistance induced by a scattering event of type $i$. In that case,

$$S_i = \frac{S_i W_i}{\sum_i W_i} \qquad (5.59)$$

And now, assuming the validity of the Wiedemann-Franz law (and forgetting the phonon contribution to thermal conductivity), one arrives at Eq. 5.57.

Another route to an identical conclusion is provided by the Mott formula. For each band,

$$S = -\frac{\pi^2}{3} \frac{k_B}{e} T \frac{\partial ln(\rho)}{\partial \epsilon} \Big|_{\epsilon = \epsilon_F} \qquad (5.60)$$

Assuming its validity for each band, this equation combined with Eq. 5.56 leads to Eq. 5.57.

Historically, the Nordheim-Gorter expression has been an attempt to separate the intrinsic Seebeck coefficient $S_0$ with the change in thermopower induced by the addition of impurities $\delta S$. If the addition of the impurities changes the resistivity from its intrinsic value $\rho_0$ by $\delta\rho$, one can write

$$S = S_i + (S_0 - S_i)\frac{\rho_0}{\rho_0 + \delta\rho} \qquad (5.61)$$

In a Nordheim-Gorter plot, the Seebeck coefficient is traced as a function of the inverse of resistivity. A linear variation is expected. The intercept and the slope of the line would yield the intrinsic $S_0$ and the impurity $S_i$, components of the Seebeck coefficient [Blatt *et al.* 1976].

The bottom line of this brief discussion is a note of caution. In real metals, the combination of the multiplicity of bands and the plurality of scattering mechanisms generates an intricate circuitry with parallel and serial channels for thermoelectricity. This is another obstacle in the way of a straightforward interpretation of the data.

# 6

# Experimental Survey I: The Periodic Table

## 6.1   Elemental Metals and Insulators

The Seebeck coefficient of a number of elements is listed in Table 6.1. The main source for the data is Macdonald [1962]. There is a bias towards metallic elements, since thermoelectricity needs mobile electrons. For comparison, the table includes also silicon, a semiconductor.

There is something remarkable about the magnitude of the thermoelectric response at room temperature. We saw in previous chapters that the Seebeck coefficient is the ratio of the entropy flow to the charge flow. Each electron has a fixed charge $e$, which is $1.6 \times 10^{-19}$ C. A reasonable guess for the amount of entropy carried by an electron would be the Boltzmann constant, $k_B = 1.38 \times 10^{-23}$ JK$^{-1}$. The ratio of these two yields $\frac{k_B}{e} = 86$ $\mu$V/K, which has the units of the Seebeck coefficient. As seen in the table, the measured Seebeck coefficient of all metals lies significantly below this value. This should mildly surprise a classical physicist, but not a quantum physicist.

Here is the door by which quantum mechanics irrupts into this picture. At room temperature, only a tiny fraction of all mobile electrons of a metal carry charge and entropy. These electrons reside very close to the interface between occupied and unoccupied states and are therefore capable of modifying their wave-vectors and participating in transport. The states available to them are restricted to a window around the Fermi energy $\epsilon_F$ with a relative width of $\frac{k_B T}{\epsilon_F}$.

The Fermi temperature of elemental metals is much larger than the ambient temperature on the surface of this planet. These solids are packed with itinerant electrons. But only a tiny fraction of these electrons participate in transport. In semimetals, on the other hand, the electron gas is much more dilute. In bismuth, the archetypical semimetal, one mobile electron is shared by $10^5$ atoms. This drastically reduces the Fermi energy, which in three dimensions scales with the carrier density to the power of two-thirds. The Fermi temperature in bismuth is of the order of room temperature. This is the fundamental reason for its Seebeck coefficient being of the order of $\frac{k_B}{e}$.

*Fundamentals of Thermoelectricity*. First Edition. Kamran Behnia.
© Kamran Behnia 2015. Published in 2015 by Oxford University Press.

Semiconductors differ radically from both metals and semimetals. Their mobile electrons exist only thanks to their thermal energy. Their concentration is much lower and decreases further as the temperature decreases. As a consequence, the electrical conductivity of semiconductors and metals differ by orders of magnitude as the reader can check by comparing the values for copper and silicon in Table 6.1. The room temperature Seebeck coefficient of silicon is significantly *larger* than $\frac{k_B}{e} = 86$ $\mu$V/K. This is because at room temperature, electrons in silicon do not follow a Dirac-Fermi distribution but a Maxwell distribution. Therefore the Fermi energy is no more a relevant energy scale and each electron has a lot of available states to choose from. This dopes the entropy flow of charged carriers.

The difference between the thermoelectric response of metals and semiconductors is such that they display opposite thermal evolutions. Upon cooling, the Seebeck coefficient decreases in metals and increases in semiconductors. The entropy-per-carrier picture of thermoelectric response provides a simple explanation. In metals, the number

**Table 6.1** *Electric resistivity and Seebeck coefficient of selected elements at 0°C n-type = electron-doped; p-type = hole-doped.*

| Element | $\rho(\mu\Omega\,cm)$ | $S(\mu V/K)$ |
|---|---|---|
| Li | 9.5 | 23.3 |
| Na | 4.9 | −5.1 |
| K | 7.5 | −11.3 |
| Rb | 13.3 | −9.4 |
| Cs | 21 | −0.8 |
| Cu | 1.7 | +1.3 |
| Ag | 1.6 | +1.2 |
| Au | 2.2 | +1.6 |
| Fe | 10 | −5.4 |
| Co | 6 | −26.7 |
| Ni | 7.2 | −16.5 |
| Pd | 10.8 | −16.6 |
| Pt | 10.6 | −12.0 |
| Pb | 21 | −0.55 |
| Bi | 130 | −108 |
| Si (n-type) | $\sim 50 \times 10^6$ | −600 |
| Si (p-type) | $\sim 50 \times 10^6$ | +500 |

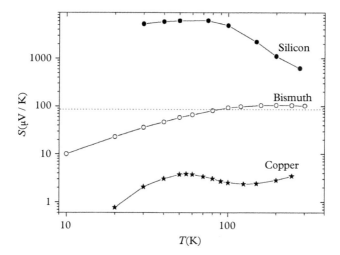

**Figure 6.1** *The Seebeck coefficient of a metal, a semimetal, and a semiconductor. Comparison of the absolute value and the temperature dependence of the Seebeck coefficient in silicon, bismuth, and copper. For bismuth, the data corresponds to the negative Seebeck coefficient measured along the trigonal axis. The horizontal dotted line corresponds to $\frac{k_B}{e} = 86 \ \mu VK^{-1}$*

of carriers is fixed and the Seebeck coefficient is set by the amount of available entropy. In a semiconductor, the available entropy is set by the energy gap, and it is the carrier concentration and its thermal evolution which determine the thermoelectric response. Figure 6.1 shows the temperature dependence of the thermoelectric response in copper, bismuth and silicon. At this level of (im)precision, the schematic picture drawn above gives a fair account of the experimental facts.

The following sections of this chapter provide a review of what is known about the thermoelectric response of a number of emblematic elements. We will see that unanswered questions abound, even in the case of these simple solids. Several families of solids have been chosen for review, each representing a category of electronic structure.

## 6.2   Alkali Metals

What can be simpler than the metals of the first column of the periodic table? Alkali atoms sit in a face-centred cubic structure, with each atom sending a single electron from its outer s-orbital to the Fermi sea. If the free-electron picture has any chance to be relevant to the real world, one should expect to find it here. In fact, both experiment and theory indicate that the Fermi surface of alkali metals is not a perfect sphere [Ham 1962]. The only alkali metal with an almost undistorted Fermi surface is sodium. In the case of all others, there is a significant departure from sphericity. In the case of caesium and most probably lithium, the distortion is large enough for a possible contact between the surface and the zone boundary.

In the early 1960s, Macdonald and collaborators [Macdonald, Pearson, and Templeton 1960, 1961] measured the Seebeck coefficient of alkali metals. The results are presented and extensively discussed in Macdonald's [1962] book. The left panel of Fig. 6.2 reproduces the data in the sub-Kelvin temperature range. As seen in the figure, even at such a low temperature, the evolution of the Seebeck coefficient with temperature is not featureless, with heavier elements displaying a stronger departure from linearity. The $T$-linear Seebeck coefficient in lithium has an embarrassingly positive sign and a slope much larger than what is expected according to its Fermi energy. Sodium, the only element whose Fermi surface is almost perfectly spherical, is also the one which displays a Seebeck coefficient of the right sign and the right amplitude. This may prove to be no coincidence.

A primary suspect for generating non-linearity is, of course, phonon drag. As seen in Fig. 6.2b, the Seebeck coefficient in both caesium and rubidium peaks below 10 K. This is reminiscent of the peak seen in noble metals at significantly higher temperatures (see Section 6.3). The Debye temperature of these solids is very low. It is 55 K in rubidium and 40 K in caesium [Macdonald, Pearson, and Templeton 1961]. The phonon drag is

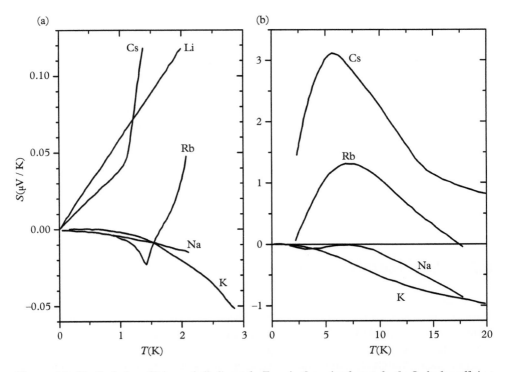

**Figure 6.2** *The Seebeck coefficients of alkali metals. Even in these simple metals, the Seebeck coefficient is not strictly linear in temperature and has an intricate structure (adapted from [Macdonald 1962]).*

expected to peak when a substantial fraction of phonons can lose their energy through scattering events involving mobile electrons. At sufficiently high temperatures, phonons are numerous and scattering among them so frequent that electron–phonon scattering events are a small fraction of total scattering events. In both alkali and noble metals, the peak attributed to phonon drag occurs at roughly a fifth of the Debye temperature.

At a low enough temperature, on the other hand, phonon drag is expected to die because of the scarcity of the electron–phonon events. In this temperature range, the typical phonon wavelength is too long for electrons to produce efficient scattering events. Since the population of phonons decrease as $T^3$, the phonon-drag contribution to the thermopower is expected to display an asymptotic $T^3$ temperature dependence and, therefore, to become negligible in comparison to the $T$-linear diffusive term. One can fit the sub-Kelvin data of Fig. 6.2 using an expression including a $T$-linear term representing the diffusive part and a $T^3$ term representing the phonon-drag component.

A more ambitious treatment has been proposed by Sondheimer [1956b], in which the effect of electron–phonon scattering on phonon energy spectrum is taken into account. Macdonald [1962] employed a simpler version of this picture and derived the following expression for phonon drag:

$$S_g = \frac{3k}{e} \left(\frac{T}{\theta}\right)^3 \int_0^{\theta/T} \frac{x^4 e^{-x} dx}{(1 + e^{-x})^2 \left(1 + \frac{T}{\Phi}\right)^3 x^3} \tag{6.1}$$

The parameter $\Phi$ quantifies the ratio of phonon–electron and phonon–electron scattering cross sections. A rather satisfactory account of the temperature dependence of the Seebeck coefficient in sodium up to 80 K was achieved [Sondheimer 1956b].

But phonon drag cannot provide a satisfactory account of the entire puzzle. We still lack a comprehensive and quantitative understanding of the thermoelectric response of various alkali metals from room temperature to below 1 K.

Most embarrassing of all is lithium. Its Seebeck coefficient is positive at room temperature. Any phonon drag contribution is expected to die upon reaching this temperature window. It is also positive and stubbornly linear in the low-temperature regime, confirming the existence of a positive diffusive thermopower even in the zero-temperature limit. Both these features are absent in sodium and potassium, which are known to have less distorted Fermi surfaces than lithium.

For a long time, the theoretically expected [Ham 1962] distortion of the Fermi surface in lithium was believed to be larger than what it actually is. It was suspected to be large enough to ensure a contact across the zone boundary, as is the case for caesium and noble metals. But this was ruled out by later studies of the Fermi surface in lithium pioneered by Donaghy and Stewart in 1967. During the next two decades, two different types of experiments led to remarkably close conclusions regarding the Fermi surface of lithium and its deviation from a perfect sphere. The Fermi-surface anisotropy for lithium is $2.8 \pm 0.6$ per cent, according to positron annihilation [Oberli *et al.* 1985], and $2.6 \pm 0.9$ per cent according to the de Haas-van Alphen effect [Randles and Springford 1976]. This is indeed greater than that for sodium or potassium, but less than that for caesium and far too little to ensure connection between neighbouring ellipsoids [Barnard 1972]. The distortion of the Fermi surface has a prominent place in speculations regarding the puzzling thermoelectric response of the noble metals.[1]

## 6.3   Noble Metals

A tight family of elements occupying a short column of the periodic table has enjoyed a place in the human history beyond what any other metal could aspire to. Noble metals are also very hard to compete with as electric conductors. These two facts are not directly linked to each other, as before the second half of the nineteenth century, electricity did not play a major role in human affairs. On the other hand, both facts are linked to other chemical and physical properties of these three elements, such as their relative stability in an oxygen atmosphere and their shininess.

The Seebeck coefficients of the noble metals (Fig. 6.3) bear additional testimony to a strong family resemblance. Both the magnitude and the temperature dependence of the thermopower, including the presence of a peak, its size, and the temperature of its occurrence, are very similar. The peak has been attributed to phonon drag and occurs, as one would expect in such a case, at a fifth of the Debye temperature. The magnitude of the room-temperature Seebeck coefficient is more or less what you would expect, given the Fermi temperature of these materials. However, instead of being negative, as one would expect for an set of electron-like carriers, it is positive.

An early consensus emerged that the peak observed near 50 K is a consequence of phonon drag. Phonon drag was expected to vanish at both high-temperature and

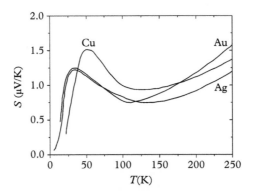

**Figure 6.3**  *The Seebeck coefficients of noble metals. A peak occurs in all three at roughly the same temperature and has been attributed to phonon drag (Adapted from [Macdonald1962]).*

---

[1] In a paper published in May 2014 (Phys.Rev. Lett. 112, 196603), Xu and Verstraete provide an explanation for the positive sign of the Seebeck coefficient invoking a non-trivial energy dependence of the scattering time.

low-temperature limits. In both these limits, electron–phonon scattering would become relatively rare. Thus, the room-temperature diffusive thermoelectricity is not affected by phonon drag. It remains puzzlingly positive in these presumably simple metals (see Fig. 6.4). This puzzle has intrigued researchers for several decades and will be discussed in more detail in Section 6.5.

Let us stress that the electronic structure of these metals is well established. Measurements of magnetic quantum oscillations (the de Haas-van Alphen effect) in the company of theoretical band calculations have led to a determination of the topology of their Fermi surfaces with impressive precision [Shoenberg 1984]; they would be spheres were they small enough to avoid contact with the boundaries of the Brillouin zone (Fig. 6.5). However, this is not the case, and adjacent spheres are interconnected across the zone boundary. As a consequence, and as seen in the right panel of Fig. 6.5, most orbits are hole-like. Such a feature has been suspected to play a role in setting the sign of the Seebeck coefficient.

Alloys of noble metals have also been a subject of investigation [Crisp and Rungis 1970]. Their thermoelectric responses provide another intriguing twist to the mystery of the positive Seebeck coefficients in these elements. Crisp and Rungis found that there is a systematic evolution of the Seebeck coefficients of silver and gold with alloying (See Fig. 6.6). Pure silver and gold present Seebeck coefficients of similar magnitude and identical sign. As they are alloyed with each other, however, a negative coefficient begins to emerge. As a consequence, both the sign and the magnitude of the room-temperature Seebeck coefficient in a alloy silver-gold at half mixing is what one would expect in the free-electron picture.

Let us now examine the available data in the zero-temperature limit. How do the free-electron expectations for thermoelectric response match with the experimental

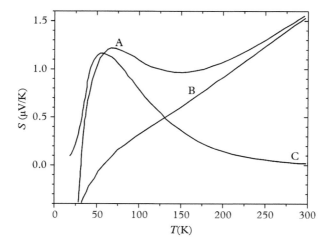

**Figure 6.4** *Diffusive and phonon-drag components of thermopower in copper. A: Total measured thermopower. B: Estimated diffusive thermopower, C: The phonon-drag component (after [Blatt, Garber, and Scott 1964]).*

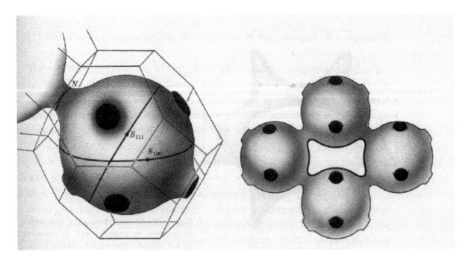

**Figure 6.5** *The Fermi surface of noble metals. Left: a set of interconnected spheres. Right: The dog-bone orbit seen by quantum oscillations is a hole-like orbit (after [Shoenberg 1984]).*

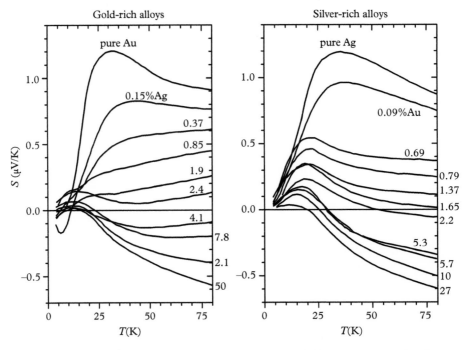

**Figure 6.6** *Seebeck coefficients in silver-gold alloys. The left (light) panel shows the data for gold-rich (silver-rich) alloys (adapted from [Crisp and Rungis 1970]).*

picture? The answer is far from straightforward and for several different reasons. While there is little uncertainty on the magnitude and the temperature dependence of the Seebeck coefficient down to 10 K, the situation starts to become more complicated in the lower temperature range. Early measurements by Gold and co-workers [Gold *et al.* 1960] on copper specimens with different impurity concentrations found strongly sample-dependent low-temperature Seebeck coefficients.

A systematic study on copper and silver was carried out several years later [Rumbo 1976]. Rumbo found that the purity of the samples, quantified by their residual resistivity at low-temperature, determines the magnitude, the structure, and even the sign of the Seebeck coefficient. Figure 6.7 shows his data on the variation of the Seebeck coefficient with sample purity. As seen in the figure, the signal shows a striking evolution with the introduction of defects. In the case of copper, the sample with an astonishing residual-resistivity ratio (RRR) of 13000 is the purest copper sample measured by Rumbo (and as far as we know anybody else). It clearly displays a negative Seebeck coefficient with a still-evolving structure.

Other authors have confirmed the strong sample dependence of the Seebeck coefficient in noble metals below 10 K. Figure 6.8 presents the data on silver and gold by

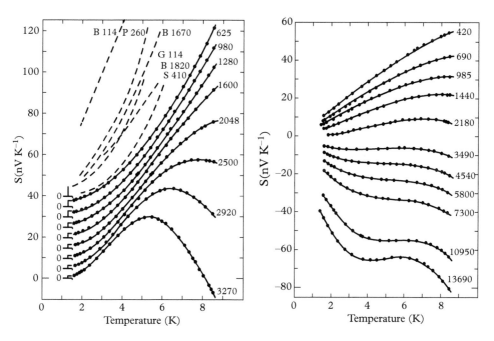

**Figure 6.7** *Seebeck coefficients in silver (left) and copper (right). Numbers indicate the ratio of the room-temperature to low-temperature resistivity (residual-resistivity ratio; RRR). In both cases, the starting sample was the purest one with the highest RRR. Defects are introduced by twisting the sample. The dashed curves are data reported by other authors. Reprinted from [Rumbo 1976], © IOP Publishing. Reproduced by permission of IOP Publishing. All rights reserved.*

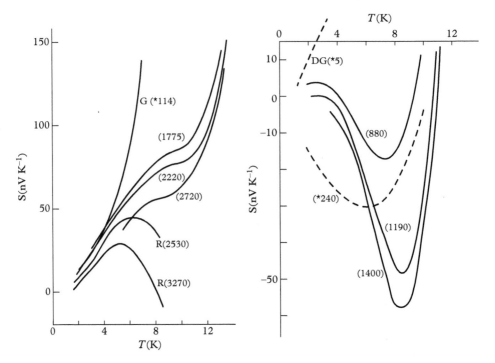

**Figure 6.8** *Low-temperature Seebeck coefficients in silver (left) and gold (right). Numbers indicate the RRR. Reprinted from [Guénault and Hawksworth 1977], © IOP Publishing. Reproduced by permission of IOP Publishing. All rights reserved.*

Guénault and Hawksworth [1977]. While there seems to be a rather broad agreement about experimental results, there has been no clear consensus on the interpretation. The phonon-drag component in silver is positive according to Rumbo and negative according to Guenault and Hawksworth and there is no obvious way to settle the issue. In such a context, neither the magnitude, nor the sign of the expected $T$-linear diffusive term in the zero-temperature limit can be determined.

At least one source of trouble complicating both the measurement and the interpretation of low-temperature thermoelectric data in metals is well established now. It is the Kondo effect. Before coming back to the long-lasting debate on the sign of the Seebeck coefficient in noble metals, let us review the Kondo effect.

## 6.4    Magnetic Impurities and the Kondo Effect

The presence of infinitesimal amount of magnetic impurities has drastic consequences on the magnitude of the low-temperature transport properties of metals. Historically, this phenomenon was recorded by the observation of an unexpected upturn in the resistivity of gold [de Haas and van den Berg 1936]. During the following decades, it became clear that *magnetic* impurities play a key role. Their presence in the host material was the origin of this upturn in resistivity.

This phenomenon has become known as the Kondo effect since the sixties. The Japanese physicist Jun Kondo treated each magnetic impurity as an isolated magnetic moment inside a Fermi sea [Kondo 1964]. Mobile electrons of the Fermi sea are scattered by these impurities through the interaction between their spin and the spin of impurities. In this process, each incoming electron visits a multitude of available states with opposite spin in the Fermi sea. Kondo found that, if the temperature is not too low, there is an elegant solution to this problem. Above a characteristic temperature, known as Kondo temperature, $T_K$, the resistivity is expected to present an additional term [Kondo 1964]:

$$\rho = \rho_i \left[ 1 + \mathcal{J}N(\epsilon_F) log\left(\frac{T}{T_K}\right)\right] \tag{6.2}$$

Here $\mathcal{J}$ is the characteristic coupling energy between two spins, the spin of the localized magnetic moment and the spin of the mobile electron. If the two spins tend to align upon interaction with each other, $\mathcal{J}$ is positive. Otherwise, it is negative. In the latter case, that is if the two spins interact antiferromagnetically, the additional resistive term of the equation mildly increases as the temperature decreases in agreement with the experiment. However, the equation forecasts a divergent resistivity at $T_K$, in contrast to the saturation observed when the sample is cooled down towards zero temperature. The perturbation approach works only for temperatures well above $T_K$. At very low temperature, the local spin of the impurity is totally screened by the spin of the mobile electrons of the Fermi sea. It is now known that such a system is a Fermi liquid with a resonant peak in the density of states near the Fermi energy.

One striking aspect of the Kondo scattering mechanism should be highlighted. When impurities are not magnetic, conduction electrons partially lose their momentum following an ordinary scattering event. This leads to a temperature-independent enhancement in resistivity. Indeed, since neither the population of impurities nor their scattering cross section change with temperature, there is no reason to expect any modification in scattering probability with cooling in this case. The situation changes when a local magnetic moment is present.

Consider a charge-carrying electron with the wave-vector $k$ scattered to the state $k'$. The electron can go through this scattering without any implication of its spin degree of freedom. However, when the coupling interaction between spins is strong enough, there is a finite probability that the electron will go from state $k$ to state $k'$ through a multiple-stage process. There is now a class of scattering events which flip the spins of both the incoming electron and the local moment. The incoming electron visits all available $k''$ states with a spin opposite to its original one before suffering another spin-flip scattering event with the impurity to end in the state $k'$ with its original spin (see Fig. 6.9). This second class of scattering events may look marginal as a way to efficiently reduce the capacity of electrons to carry charge. Indeed, the most drastic consequence

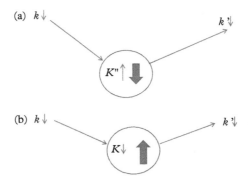

**Figure 6.9** *Scattering with (a) and without (b)
spin flip. As the temperature is reduced towards
the Kondo temperature, an increasing
proportion of scattering events flip the spins of
the incoming electron and the localized moment.
This is the origin of the upturn in resistivity.*

of putting iron impurities in gold is to increase the temperature-independent residual resistivity, which occurs upon the introduction of any kind of impurity.

At the heart of Kondo's solution to the mystery of the resistivity upturn resides the fact that the relative weight of the double-spin-flip process of Fig. 6.9 increases as the temperature is reduced and the thermal distribution of the electrons near the Fermi energy sharpens. As the temperature decreases, an increasing fraction of scattering events occurs like this. When $\mathcal{J}$ is negative, this kind of scattering is more efficient than ordinary scattering in impeding charge conductivity; hence the upturn in resistivity.

The Kondo effect leads to a spectacular increase in the low-temperature Seebeck coefficients of noble metals. The effect has been extensively documented by Macdonald and collaborators [Macdonald 1962; Macdonald, Pearson, and Templeton 1962]. In 1965, Kondo considered the thermoelectric response in the context of magnetic-impurity scattering of conduction electrons. Following an empirical equation communicated by Guénault, he proposed a simple expression for the thermopower of a Kondo system:

$$S = A\frac{k_B}{e}\frac{T}{T + T_0} \tag{6.3}$$

He argued that the thermopower becomes large and almost temperature independent well above the characteristic temperature, $T_0$. In this temperature window, it is of the order of $A\,\frac{k_B}{e}$, where $A$ is a numerical coefficient somewhat less than unity. In this regime, the Seebeck coefficient is set by the very large energy-dependent scattering rate. This energy derivative increases as the temperature is reduced and cancels the reduction generated by the thermal width of the configuration entropy $k_B T$. Below $T_0$, the large Seebeck coefficient is expected to decrease linearly with temperature, giving rise

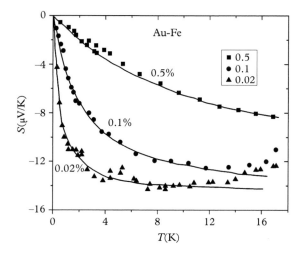

**Figure 6.10** *The effect of iron impurities on the Seebeck coefficient of gold. Symbols are experimental data points [Macdonald, Pearson, and Templeton 1962] and the lines represent theoretical expectation [Kondo 1965].*

to the asymptotic $T$-linear dependence expected by the equation. As seen in Fig. 6.10, the simple equation gives a rather satisfactory account of the early experimental data. It remains a very crude phenomenological description of the physical phenomena.

More refined treatments of the Seebeck coefficient of a Kondo system were carried out by Fischer [1967] and Maki [1969]. A key parameter in these approaches is the phase shift of the electron wave-function induced by scattering. The final results of these calculations are impressively elaborate and difficult to be unambiguously verified by experiment.

On the experimental side, the Seebeck coefficients of Kondo systems have been investigated with a variety of magnetic impurities inserted in different host materials, which were often noble metals. A detailed list can be found in the book by Blatt and co-workers [Blatt *et al.* 1976]. The case of iron impurities in a gold matrix has been extensively documented [Berman and Kopp 1971; Kopp 1975; Chaussy *et al.* 1982]. In particular, Kopp [1975] measured samples with an extremely low level of iron impurities. As seen in Fig. 6.11, even concentrations as small as one part in a billion have a detectable signature in the Seebeck coefficient.

The manyfold enhancement of the Seebeck coefficient induced by the Kondo effect is remarkable. It is to be contrasted with a very modest (typically a few parts out of a hundred) enhancement induced in resistivity. Let us recall that electrons which participate in charge transport are mostly those which are exactly at the Fermi level. Thermal broadening allows a residual fraction of charge-carrying electrons to be slightly below or slightly above the Fermi level. For thermal and thermoelectric transport on the other hand, the situation is different. Electrons which are exactly at the Fermi level are exempt

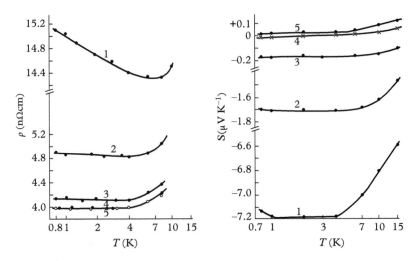

**Figure 6.11** *The effect of iron impurities on low-temperature resistivity (left) and Seebeck coefficient (right) of gold. Iron concentration is 13 ppm in sample 1; 1 ppm in sample 2; 0.1 ppm in sample 3; 0.01 ppm in sample 4; and 0.001 ppm in sample 5. Only the purest sample (with an RRR of about 400) presents a positive Seebeck coefficient. Reprinted from [Kopp 1975], © IOP Publishing. Reproduced by permission of IOP Publishing. All rights reserved.*

from carrying entropy. The entropy-carrying electrons are those which are slightly above or slightly below the Fermi level. Moreover, the contributions from electrons below and above the Fermi level have opposite signs and thus cancel each other out. Now, the Kondo physics implies a resonant energy scale slightly below the average energy of the typical member of the Fermi sea. Its effect on resistivity is modest since it concerns a small fraction of charge-carrying electrons. On the other hand, a lot of entropy can be carried by electrons which happen to be at the right side of the Fermi level. This is the fundamental reason behind the extreme sensitivity of the Seebeck coefficient (see Fig. 6.12).

Interestingly, the low-temperature, thermopower-enhanced Seebeck coefficient induced by the Kondo effect occurs in a very limited window of iron concentration inserted in gold. This can be seen in Fig. 6.11. Putting together the data reported by several groups, one can see that helium-temperature thermopower is enhanced when the iron concentration is somewhere between 100 and 700 ppm (see Fig. 6.12). Within these limits, the iron–gold system has provided us with a most sensitive thermocouple component. Below this concentration, the Kondo effect gradually dies away. Above this concentration, magnetic impurities are dense enough to affect the spin of each other. The system begins to become a spin-glass system. In this regime, isolated spins are no longer screened by conduction electrons, and the entropy carried by conduction electrons drastically diminishes. In the other extreme, that is, at the dilute limit, one expects to recover the diffusive thermoelectricity of the pure system.

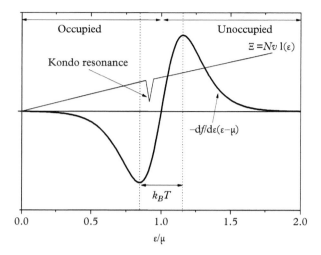

**Figure 6.12** *Kondo resonance leads to a drastic imbalance between occupied and unoccupied states.*

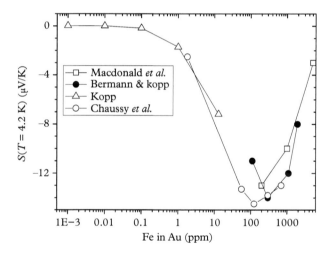

**Figure 6.13** *Evolution of thermoelectric power of gold–iron alloys at 4.2 K as a function of iron impurities. Based on the data reported by [Macdonald et al. 1962, Berman & Kopp 1971, Kopp 1975 and Chaussy et al. 1982].*

This brings us back to the question of the diffusive thermopower in noble metals at very low temperatures. In the case of iron impurities in copper, the early data by Gold and co-workers [Gold *et al.* 1960] do not permit us to establish the existence of such a restricted concentration window for enhanced thermoelectric response. As seen in Fig. 6.11, the purest sample studied by Kopp shows a positive Seebeck coefficient. Kopp concluded that this positive signal represents the diffusive thermoelectric response of pure gold. Interestingly, the magnitude of the $T$-linear term for this sample ($+7$ nV K$^{-2}$) is close to that yielded by the room-temperature Seebeck coefficient for gold (see Fig. 6.3).

Does this mean that the case of the amplitude (and sign) of the diffusive Seebeck coefficient in gold is settled? The answer is no. As seen in Fig. 6.11, the residual resistivity of this sample is 4 n$\Omega$ cm. This corresponds to an RRR of 600 (the room-temperature resistivity of gold is 2.4 $\mu\Omega$ cm), which is well below the purity reached by Guénault and Rumbo, whose data were discussed in Section 6.3 (Fig. 6.7 and Fig. 6.8). As we saw there, upon further increase in RRR, the low-temperature Seebeck coefficients of all three noble metals change sign again to become negative.

Thus, on the experimental plane, the asymptotic zero-temperature diffusive thermopower of pure gold remains an open question. Neither the magnitude nor the sign of the Seebeck coefficient of the intrinsic material can be considered to be set beyond reasonable doubt.

## 6.5   Origin of the Positive Seebeck Coefficients of Noble Metals

*(For more than thirty years the absolute thermoelectric power of pure samples of monovalent metals has remained a nagging embarrassment to the theory of the ordinary electronic transport properties of solids. All familiar simple theory has promised us that in these materials the sign of the electron-diffusion contribution to the thermopower should be that of the charge carriers as determined by the Hall effect, i.e. negative; but instead it turns out to be positive for Cu, Ag, Au and—even more perversely—for Li alone of the solid alkalies. At least two generations of experimentalists have remained completely unshaken in testifying to these results as obstinate facts of life.)*

These are the opening sentences of a 1967 paper by John E. Robinson, who presented a simple solution to this puzzle [Robinson 1967]. Mysteriously, decades later, both the puzzle and the solution he proposed are widely forgotten. The focus of contemporary condensed-matter physics is elsewhere. The puzzle of thermoelectricity in noble metals becomes more striking when one considers that the intricate details of the Fermi surface of these materials are well established and this knowledge is a testimony to the oft celebrated glory of the band theory of metals.

The electronic specific heat of noble metals has also been measured with great precision [Martin 1973]. Intriguingly, $q$, the dimensionless ratio of thermopower to the electronic specific heat (defined in Section 5.8), is close to unity for all these three metals. (See Table 6.2). This confirms that we are indeed facing diffusive thermoelectricity of the right magnitude but with the wrong sign and looking like what one would expect for a free gas of holes! But this cannot be.

The solutions proposed for this puzzle fall in to several categories. The first are those which highlight the shape of the Fermi surface in noble metals (Fig. 6.5); this shape is not spherical and touches the Brillouin zone. The sign and the magnitude of the Seebeck coefficient is set by the average contribution of electrons distributed over the whole Fermi

**Table 6.2** *The slope of the room-temperature Seebeck coefficient, the electronic specific heat of noble metals, and the dimensionless ratio of the two, $q = \frac{SN_{Av}e}{T\gamma}$.*

|  | Cu | Ag | Au |
|---|---|---|---|
| $S/T \, (\text{nV K}^{-2})$ | +5.4 | +4.8 | +6.2 |
| $\gamma \, (\mu\text{J K}^{-2} \text{mol}^{-1})$ | 691 | 640 | 689 |
| $q$ | +0.75 | +0.81 | +0.86 |

surface, which has both electron-like and hole-like curvatures. If one could find a way to show that the contribution of the small 'neck' can outweigh that of the large 'belly', then a positive Seebeck coefficient in such an electron Fermi surface is conceivable. This tempting scenario was considered by Ziman, who noticed that this scenario has to confront the stubborn fact that the Hall coefficient of these metals is negative and its magnitude is very close to what is expected from the free-electron picture [Ziman 1961].

The strongest experimental evidence *against* any major role played by the distorted shape of the Fermi surface is the fact that molten noble metals continue to display positive Seebeck coefficients. It is hard to imagine that the Fermi surface of the liquid state would be anything but a sphere. Another is the negative Seebeck coefficients for alloys of two noble metals (see Section 6.3). Since the Fermi-surface topology is barely different between the two metals, one would expect the alloy to possess a Fermi surface of similar distorted shape. Why then does the Seebeck coefficient becomes negative?

In another category is Robinson's proposed solution, which does not need a distorted Fermi surface. He recalled that the Seebeck coefficient of a crystal, irrespective of the shape and the curvature of its Fermi surface, is set by the outcome of two contributions of opposite signs: one from occupied states ('electrons') and one from unoccupied states ('holes'). If the mean-free-path is constant, then unoccupied states outweigh occupied ones, because both the velocity and the density of states increase with increasing energy. In this case, the sign of the diffusive thermopower is the same as the Hall coefficient (see Fig. 6.14a). But what if the mean-free-path decreases with increasing energy? If this feature happens to be strong enough, it can invert the balance and modify the expected sign of the Seebeck coefficient (Fig. 6.14b).

This statement can be quantified using Eq. 5.25. The sign of the Seebeck coefficient will become opposite to the one expected for the constant mean-free-path case, if the following inequality holds:

$$\frac{\partial ln(\ell)}{\partial ln(\epsilon)}\Big|_{\epsilon=\epsilon_F} < -1 \tag{6.4}$$

The simplest transport theory assumes that the mean-free-path is just the distance between two scattering impurities and thus constant. Another common simple assumption is a constant relaxation time between two scattering events. Since in the latter case, the faster electrons travel longer before being scattered and thus $\frac{\partial ln(\ell)}{\partial ln(\epsilon)}\Big|_{\epsilon=\epsilon_F}$ would be

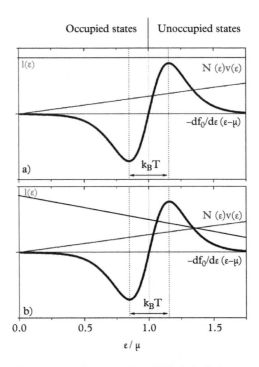

Occupied states | Unoccupied states

**Figure 6.14** *The sign of the Seebeck coefficient is set by the product of three functions. (a) In the simplest case, the mean-free-path is energy independent. When the dispersion is parabolic, electrons with higher energy have a larger velocity and density of states than those with lower energies and thus they contribute more to the Seebeck coefficient. The sign of the Seebeck coefficient is set by them. (b) But the mean-free-path may present a strong energy dependence. If the mean-free-path of more energetic electrons happens to be much shorter, then the final sign of the Seebeck coefficient can be reversed, even for the same topology of Fermi surface.*

positive. In order to find a $q$ close to unity with an inverted sign, $\frac{\partial ln(\ell)}{\partial ln(\epsilon)}\big|_{\epsilon=\epsilon_F}$ should be negative and large.

Bourassa, Wang, and Lengeler [Bourassa 1978] have scrutinized the fine structure of the Fermi surface of the noble metals obtained by de Haas-van Alphen measurements to obtain the magnitude of negative $\frac{\partial ln(\ell)}{\partial ln(\epsilon)}\big|_{\epsilon=\epsilon_F}$ required to quantitatively explain the positive Seebeck coefficients of noble metals. The diffusive Seebeck coefficient can be written as

$$S = -\frac{\pi^2}{3} \frac{k_B}{e} \frac{T}{T_F} \left[ \frac{\partial ln(A(\epsilon))}{\partial ln(\epsilon)} \Big|_{\epsilon=\epsilon_F} + \frac{\partial ln(\ell(\epsilon))}{\partial ln(\epsilon)} \Big|_{\epsilon=\epsilon_F} \right] \qquad (6.5)$$

Here, $A(\epsilon)$ is the area of the constant energy $\epsilon$. Its energy derivative at the Fermi level is given by the de Haas-van Alphen data on the curvature of the Fermi surface of noble metals [Shoenberg 1984]. Bourassa and co-workers used these data together with the Fermi temperature derived from band calculations to extract $\frac{\partial ln(\ell)}{\partial ln(\epsilon)} \Big|_{\epsilon=\epsilon_F}$ for each of the three noble metals. The results are given in Table 6.3.

An alternative method is to use the specific heat data. Assuming that both the Fermi temperature and the energy derivative of the constant energy are contained in electronic specific heat, one can write

$$q = -\frac{2}{3} \left[ 1 + \frac{\partial ln(\ell(\epsilon))}{\partial ln(\epsilon)} \Big|_{\epsilon=\epsilon_F} \right] \qquad (6.6)$$

As seen in Table 6.3, both routes arrive to comparable values. Assuming that room-temperature Seebeck coefficients of noble metals are diffusive and the thermodynamic contribution is positive, large negative values of $\frac{\partial ln(\ell)}{\partial ln(\epsilon)} \Big|_{\epsilon=\epsilon_F}$ are required to explain the data. Is there a realistic way to do this?

Robinson argued that the answer to this question is positive and presented a model of electron–phonon scattering leading to a drastically energy-dependent mean-free-path [Robinson 1967]. In the model he proposed, the scattering potential between electrons and phonons depends on the wave-vector exchanged during the interaction, $q$, in the following way:

$$V(q) = \frac{D - ne^2/q^2}{1 + k_s^2/q^2} \qquad (6.7)$$

The two terms in the nominator represent the bare Coulomb potential between electrons and ions. To the attractive potential $ne^2/q^2$, a constant term $D$, which is a deformation potential arising from the charge necessary to keep the chemical potentials constant, is added. The denominator represents the screening term, with $k_s$ being the Thomas-Fermi screening wave-vector. Now, since such a scattering potential vanishes for finite $q$, the scattering cross section should exhibit a minimum as a function of energy. On the high-energy side of this minimum, the mean-free-path of electrons will decrease with increasing energy. Robinson quantified this argument by carrying out

**Table 6.3** *Values of $\frac{\partial ln(\ell)}{\partial ln(\epsilon)} \Big|_{\epsilon=\epsilon_F}$ for noble metals using the de Haas van Alphen (dHvA) or the specific heat data.*

|               | Cu    | Ag    | Au    |
|---------------|-------|-------|-------|
| dHvA          | −2.07 | −1.61 | −2.14 |
| Specific heat | −2.12 | −2.21 | −2.29 |

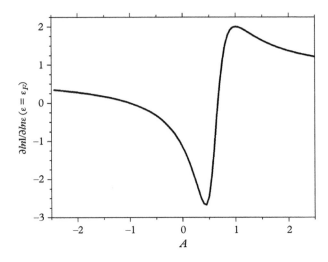

**Figure 6.15** *Variation of the logarithmic energy derivative of the mean-free-path as a function of parameter A according to a model of electron–phonon interaction proposed by Robinson [Robinson 1967]. A quantifies the relative strength of screening and ion core potentials. In a narrow window, the mean-free-path of electrons becomes a sharply decreasing function of their energy.*

detailed calculations for a spherical Fermi surface [Robinson 1967]. Figure 6.15 shows one of his typical curves of $\frac{\partial ln(\ell)}{\partial ln(\epsilon)}|_{\epsilon=\epsilon_F}$ as a function of $A$, defined below:

$$A = \frac{ne^2}{k_s^2 D} \tag{6.8}$$

As seen in the figure, in a narrow window, when $0 < A < 1$, the magnitude of $\frac{\partial ln(\ell)}{\partial ln(\epsilon)}|_{\epsilon=\epsilon_F}$ drastically changes. This is where the screening potential and the ion bare potential are of comparable magnitudes. As seen in the figure, it can become negative and large enough ($\sim -3$) to make this scenario a credible explanation of the positive sign of the Seebeck coefficient in noble medals. However, there has been no independent experimental confirmation of such a drastic energy-dependent mean-free-path in noble metals.

A third possible solution to the puzzle of the positive Seebeck coefficients in noble metals was proposed recently [Sonntag 2010]. The treatment of diffusive thermoelectricity assumes rigid bands, which do not shift as a function of temperature. However, in real solids, atomic distances vary as their temperatures is raised and the band edges shift with temperatures. Sonntag argued that if the shift in the band edge is strong enough (of the order of 25 $\mu$eV K$^{-1}$ in copper), the effect is strong enough to change the sign of the Seebeck coefficient. Such a small shift cannot be excluded. It would mean that the Fermi energy at zero temperature and room temperature differ by $10^{-3}$. This scenario is yet to be experimentally tested.

Both of the last two proposed solutions have no difficulty extending their explanation to the liquid phase of noble metals. However, there are two other pieces of experimental puzzle, which can only be accounted for by one of the two scenarios.

As we saw in the previous section, the positive Seebeck coefficient vanishes in an silver–gold alloy, which has a negative room-temperature Seebeck coefficient comparable in magnitude to that of pure silver or pure gold. The solid alloy has presumably the same Fermi-surface topology and the same temperature-induced band edge as the two pure constituents (since they barely differ between silver and gold). On the other hand, the strongly energy-dependent mean-free-path scenario can easily confront this experimental fact. The electron mean-fee path is drastically reduced in the alloy, and scattering is essentially elastic. In the absence of inelastic scattering, the alloy is expected to recover its negative diffusive thermopower.

The other piece of the puzzle is the strong dependence of the Seebeck coefficient below 1 K on the defect concentration (Fig. 6.7 and Fig. 6.8). In this temperature range, the band edge can only be identical to its zero-temperature value. The strong dependence of the Seebeck coefficient on defect concentration indicates that it is extremely sensitive to the mean-free-path. However, in this temperature range, phonons are not expected to play a major role in scattering and what is required is an energy-dependent electron-impurity scattering mechanism.

In summary, at this stage, the energy dependence of the electronic mean-free-path remains the most promising of the proposed solutions to the puzzle of the positive Seebeck coefficients in noble metals and lithium (see [Robinson and Dow 1968]). However, an independent experimental confirmation of the large negative magnitude of $\frac{\partial ln(\ell)}{\partial ln(\epsilon)}|_{\epsilon=\epsilon_F}$ postulated in this scenario has not emerged yet.

## 6.6 Column V Semimetals

Semimetals are metals with a small concentration of carriers of both signs. The two emblematic elemental semimetals are graphite and bismuth. In both these elements, the concentration of mobile carriers is so low that the electron fluid becomes much more dilute than the one found in metals. In bismuth, there are $3 \times 10^{17}$ electrons in a cubic centimetre. There is also an identical concentration of hole-like carriers. Such a concentration implies that, instead of having roughly one electron per atom as in the case of a noble (or any other ordinary) metal, a single mobile electron is shared by $10^5$ atoms. The carrier density is one order of magnitude larger ($4 \times 10^{18}$ cm$^{-3}$) in graphite and almost two orders of magnitude larger in antimony [Issi 1979].

The first measurements of the Seebeck coefficient in bismuth were carried out by Gallo and co-workers from 80 K to room temperature. The Seebeck coefficient was found to be large, negative, and anisotropic [Gallo *et al.* 1963]. The data is reproduced in Fig. 6.16. When these measurements were performed, the structure of the Fermi was not fully known. During the following years, thanks to a number of de Haas-van Alphen studies, the number, shape and orientation of all Fermi surface pockets in bismuth were established in great detail [Bhargava 1967; Shoenberg 1984].

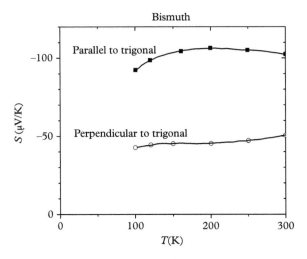

**Figure 6.16** *First data on the Seebeck coefficient of bismuth between 80 K and 300 K (adapted from [Gallo et al. 1963]).*

It consists of a hole pocket and three electron pockets (see [Issi 1979] for a review). The hole pocket is as large as the sum of the three electron pockets, which lie in a plane almost perpendicular to the longer axis of the hole pocket. The electron pockets, located at the L-point of the Brillouin zone, have a non-parabolic dispersion. This feature has attracted recent attention in the context of research on Dirac electrons in condensed-matter physics. The surprisingly complex Landau spectrum of the system at high magnetic fields can be understood on the basis of this Fermi-surface topology [Zhu *et al.* 2011].

Based on the knowledge that bismuth is a compensated semimetal, Gallo and coworkers provided satisfactory explanations for both the sign and the magnitude of the Seebeck coefficient. It is large because the Fermi energy in bismuth is low (see Table 6.4). It is negative because there are multiple pockets of electrons and a single pocket of holes, with comparable Fermi energies. Moreover, electrons are more mobile than holes; thus, they have a larger conductivity, so their contribution to the Seebeck coefficient outweighs the contribution of holes.

The anisotropy of the Seebeck coefficient can be explained by the particular arrangement of the hole and the electron ellipsoids in bismuth's Fermi surface (Fig. 6.17).

**Table 6.4** *Carrier density and Fermi energies of Column V semimetals (after [Issi 1979]).*

|  | Bi | Sb | As |
|---|---|---|---|
| carrier density ($10^{17}$ cm$^{-3}$) | 2.7 | 37.4 | 200 |
| Fermi energy (meV) | 11–27 | 84–93 | 21–202 |

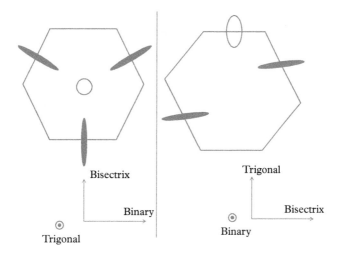

**Figure 6.17** *The Fermi surface in bismuth consists of one hole ellipsoid and three electron ellipsoids lying in planes almost perpendicular to each other.*

The Seebeck coefficient along the trigonal axis, $S_{\parallel}$, and that perpendicular to it, $S_{\perp}$, can be expressed as

$$S_{\parallel} = \frac{1}{\sigma_{\parallel}} \left[ S^h \sigma^h_{\parallel} + \sum_i S^{ei} \sigma^{ei}_{\parallel} \right] \tag{6.9}$$

$$S_{\perp} = \frac{1}{\sigma_{\perp}} \left[ S^h \sigma^h_{\perp} + \sum_i S^{ei} \sigma^{ei}_{\perp} \right] \tag{6.10}$$

We are assuming the same isotropic Seebeck coefficient for holes and electrons. This assumption is valid if the magnitude of Seebeck coefficient of each ellipsoid is set by its Fermi energy. In this case, the anisotropy arises because of the anisotropy in charge conductivity. Now, the longer axis of the hole ellipsoid is parallel to the trigonal axis. Thus,

$$\sigma^h_{\parallel} < \sigma^h_{\perp} \tag{6.11}$$

Because the electron pocket's longer axis is almost perpendicular to the Fermi surface, the inequality is reversed in the case of electrons:

$$\sigma^e_{\parallel} > \sigma^e_{\perp} \tag{6.12}$$

We can now see why the absolute value of the Seebeck coefficient is two times larger when the thermal gradient is along the trigonal axis. Electrons conduct better and holes conduct less in this configuration. The difference between the contributions of electrons and holes is greatest in this configuration. When the thermal gradient is perpendicular to the trigonal axis, the difference between the hole and electron contribution is much less since the holes' conductivity is maximal and electron conductivity is minimal.

The almost temperature-independent Seebeck coefficient in bismuth can also be qualitatively understood. The gap between conduction and valence bands in bismuth is small. When the temperature is comparable or larger than this gap, the carrier population increases with increasing temperature, pushing up the Fermi temperature (for both electrons and holes). The temperature-induced decrease in the Fermi temperature is such that the ratio $\frac{T}{T_F}$ does not vary much above 100 K, leading to a flat Seebeck coefficient.

During the two decades following the original measurements by Gallo and co-workers, both the measurements and the analysis were extended by several authors [Korenblit *et al.* 1969; Boxus and Issi 1977; Uher and Pratt 1978]. A comprehensive review was published by Issi [1979], who compiled the data for all semimetallic elements of Column V (Fig. 6.18). As seen in the figure, the Seebeck coefficient for bismuth below 100 K begins to decrease towards zero before showing a phonon-drag peak close to 4 K, with opposite signs for the two orientations of the thermal gradient. The lattice thermal conductivity in bismuth peaks almost at the same temperature. Korenblit, Kusnetsov, and Shalyt [1969] argued that, in a semimetal such as bismuth, carriers are confined to their tiny pocket, which is located at the boundary of the Brillouin zone and far apart from other pockets. At low temperatures, it costs too much impulsion to scatter an electron from one pocket to the other. In such a context, only phonons with a long

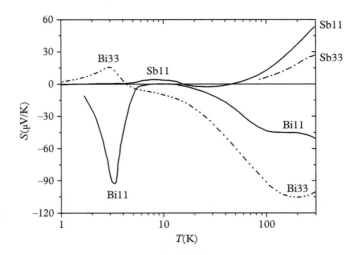

**Figure 6.18** *Seebeck coefficients of bismuth and antimony. The numbers 33 and 11 refer to a thermal gradient applied along (perpendicular) to the trigonal axis (adapted from [Issi 1979]).*

wavelength (and a small wave-vector) can couple to electrons. Now, as the temperature is reduced, the wave-vector of an acoustic phonon with thermal energy decreases, with a slope set by the sound velocity $v_s$.

$$\hbar k_s \simeq \frac{k_B T}{v_s} \tag{6.13}$$

At low enough temperatures, $k_{ph}^{th}$ becomes comparable to the average diameter of an electron pocket, $2 < k_F >$. This sets a new temperature scale for the most favourable conditions for momentum exchange between electrons and phonons:

$$T_0 \simeq \frac{2\hbar < k_F > v_s}{k_B} \tag{6.14}$$

Indeed, injecting appropriate parameters in to this equation yields a $T_0$ of the order of a few kelvins, close to the temperature at which the phonon drag peak is observed. Several studies [Issi and Mangez 1972; Boxus and Issi 1977; Boxus et al. 1981] have found that the amplitude of the phonon-drag peak is set by the phonon mean-free-path. Experiments on samples with different dimensions provide compelling evidence for this. The smaller the sample, the lower the maximum lattice thermal conductivity and the smaller the phonon-drag peak. Thus, there is a qualitative understanding of several features of the phonon-drag peak in bismuth. However, a quantitative description has remained both elusive and a subject of controversy [Jacobson and Ertl 1972].

Above 20 K, the Seebeck coefficient remains mainly diffusive, and its detailed description remains a formidable challenge. One particularity of bismuth is the smallness of the energy gap between conduction and valence bands at the L-point of the Brollouin zone, where the electron pockets reside. There are two important consequences for any analysis of the thermoelectric response. First, the energy dispersion of the carriers in the conduction band is no longer parabolic and now contains a linear term. The appropriate Hamiltonian to describe the carriers in bismuth at L-point is akin to the Dirac Hamiltonian [Wolff 1964].

There is a second outcome to the small magnitude of the energy gap. As the temperature becomes comparable to the energy gap, thermally excited carriers begin to populate the conduction band, and the carrier concentration steadily increases with the increasing temperature. As a consequence, the Fermi energy at finite temperature substantially differs from its zero-temperature value. Since the system is compensated and the number of electrons and holes are to remain equal, this is true for both electrons and holes.

A quantitative description of the temperature dependence of the Seebeck coefficient in bismuth should take into account both the non-parabolicity of the energy dispersion for electrons and the steady shift of the Fermi temperature for both electrons and holes. Reconciling these with the experimental data has proved to be a very difficult task [Hansen et al. 1978; Heremans and Hansen 1979]. Ironically, above 100 K,

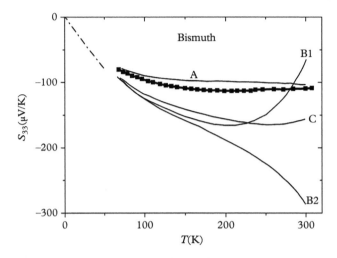

**Figure 6.19** *Seebeck coefficients in bismuth and theoretical models. The thick solid line with solid squares represents the experimental data. Other solid lines correspond to different theoretical attempts to describe this data. Curve A is based on a parabolic model. B1, B2, and C are expectations of three different versions of a non-parabolic model with different assumptions (after [Heremans and Hansen 1979]).*

when the band parameters shift with temperature, it is easier to fit the data neglecting the non-parabolic dispersion (see Fig. 6.19). Thus, a satisfactory quantitative description of the thermoelectric response in bismuth close to room temperature is yet to be attained.

The Nernst coefficient in bismuth in the low-field limit has been the subject of several studies [Korenblit, Kusnetsov, and Shalyt 1969]; Sugihara 1969; Behnia, Méasson, and Kopelevich 2007a]. It shows a complex temperature dependence that is yet to be quantitatively understood. There is a substantial phonon-drag component in the Nernst response. This is indicated by the fact that it peaks at a temperature close to the maximum of the Seebeck coefficient and the lattice thermal conductivity. There, the diffusive component to the Nernst response is also very large, a consequence of the very high mobility of carriers. It easily dwarfs what has been seen in other metals. Thus, the solid in which Nernst and Ettingshausen discovered the effect bearing their name is still the metal known to present the largest Nernst coefficient [Behnia 2009].

The two other semimetals of Column V (arsenic and antimony) share the crystal structure of bismuth. But, they have a larger concentration of mobile electrons. As seen in Fig 6.18, the Seebeck coefficient of antimony is also anisotropic, albeit with a lower magnitude than that for bismuth. This is not surprising, given antimony larger Fermi energy, the primary parameter to set the magnitude of thermoelectric response. It is also remarkable that the phonon drag peaks for arsenic and antimony occur at higher temperatures. Since, as one moves upwards in the column, the Fermi surface becomes larger in diameter, the scenario put forward by Korenblit and co-workers, briefly mentioned above, is compatible with this feature.

Alloys of bismuth and antimony are known to be interesting thermoelectric materials because of their sizeable figures of merit at cryogenic temperatures. We will review these alloys along other narrow-gap semiconductors in Chapter 7.

## 6.7   Column IV Semiconductors

Column IV of the periodic table is host to diamond, silicon, and germanium. These are insulators with a band gap which, as one moves downwards in the column, decreases in magnitude. Arguably, silicon is the most technologically important element of our era. It comes as a surprise, therefore, to find out that only a few studies devoted to its thermoelectric properties can be found in scientific literature.

In the 1950s, Geballe and Hull published two papers reporting their measurements of the Seebeck coefficients in germanium [Geballe and Hull 1954] and in silicon [Geballe and Hull 1955]. Their data extend from 20 K to slightly above room temperature and covers a wide range of both n- and p-doping in both systems. More than half a century later, these two papers remain the main source of our information on the thermoelectric response of the two archetypical semiconductors. Only a handful of other and less extensive reports on this subject is known to the author of these lines.

Fulkerson and co-workers studied the Seebeck coefficient of two silicon samples without added dopants up to 1300 K [Fulkerson *et al.* 1968]. Weber and Gmelin reported on transport properties of two single-crystalline silicon samples, one lightly and the other heavily doped [Weber and Gmelin 1991]. Their data extend from room temperature to 4.2 K. Finally, In the 1990s, Löhneysen and his collaborators [Lakner and Löhneysen 1993; Liu *et al.* 1996; Löhneysen 2011] studied the Seebeck coefficient of doped silicon in the vicinity of the metal–insulator transition at low temperatures (the temperature window of the data extends well below 1 K). To this short list, one may add two studies of Seebeck coefficients in germanium cooled down to liquid helium temperatures [Mooser and Woods 1955; Kaden and Günter 1984].

Geballe and Hull presented their data (Fig. 6.20) in a manner unfamiliar to a contemporary reader of reports on such measurements. What is plotted is the product of the Seebeck coefficient and absolute temperature. According to the Kelvin relation, this product is nothing else than the Peltier coefficient. The latter is the ratio of heat flow to the charge flow causing it. In other words, it measures the thermal energy each flowing electron carries.

In a metal, this ratio is the difference between the chemical potential, the energy cost of adding one electron to the system, and the Fermi energy. As the temperature rises, the chemical potential shifts away from the Fermi energy with a quadratic temperature dependence, leading to a $T$-square Peltier coefficient and a $T$-linear Seebeck coefficient.

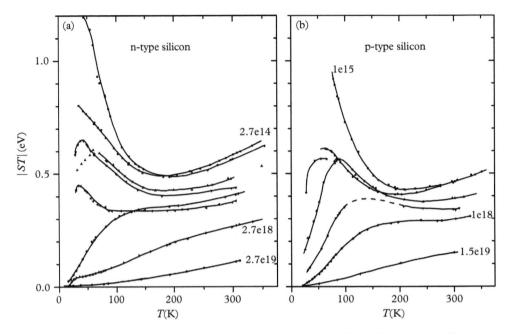

**Figure 6.20** *The Peltier coefficient (ST) in electron-doped (n-type) and hole-doped (p-type) silicon. Different curves refer to samples with different concentrations of various dopants. The carrier concentration in most and least doped samples is specified (adapted from [Geballe and Hull 1954]).*

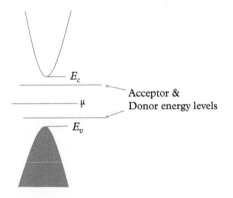

**Figure 6.21** *Energy scales in a semiconductor.*

In a semiconductor, there is an (upper) conduction band (which is empty) on top of a full (lower) valence band (Fig. 6.21). The chemical potential is between the two, inside the gap separating them. At zero temperature, carriers are frozen and no charge flows in this system. At finite temperatures, charge can flow, thanks to thermally excited carriers. The thermal energy of each of these carriers is the distance between the chemical potential and the top of the valence band, $E_v$ (or the bottom of the conduction band, $E_c$). In other words, one would expect a Peltier coefficient for electrons of the order of

$$\Pi_e \approx \frac{(E_c - \mu)}{-e} \tag{6.15}$$

As for holes, the same line of argument leads to

$$\Pi_h \approx \frac{(E_v - \mu)}{e} \tag{6.16}$$

Given that the chemical potential is roughly midway between $E_c$ and $E_v$, the order of magnitude suggested by these expressions for the Peltier coefficient is half of the band gap, $\Delta = E_c - E_v$. With this primitive approximation in mind, the numbers in Fig. 6.20 should be compared with the magnitude of the band gap in silicon, which is 1.1 eV.

Now, let us note that, in an intrinsic semiconductor with a perfectly symmetric valence and conduction bands, the hole and the electron contribution to the Seebeck effect would cancel out. But, this is almost never the case. There is a dissymmetry between the two, generating a finite thermoelectric response. In the case of silicon in particular, the energy gap is indirect, and the bottom of the conduction band and the top of the valence band are not located at the same point in the $k$-space. It is not surprising, therefore, that holes and electrons in silicon differ in their effective mass and their mobility. The contribution of each type of carriers to the total Seebeck coefficient depends on their respective mobilities:

$$S_{total}\sigma_{total} = S_h\sigma_h + S_e\sigma_e = e(|S_h|n_h\mu_h - |S_e|n_e\mu_e) \tag{6.17}$$

Thus, even when there are equal concentrations of thermally excited electrons and holes ($n_h = n_e$), the difference in mobility leads to a finite thermoelectric response and determines its sign. The product $ST$ is expected to stay smaller than half of the energy gap, $\frac{\Delta}{2}$.

The discussion above assumed a perfect and pure semiconducting crystal. In the real world, additional energy scales are introduced by the presence of extrinsic dopants or simply by atomic vacancies. The system can be doped either by n-type donors, which generate a population of electrons with an energy slightly larger than the top of the valence band or by p-type acceptors creating an energy level just below the conduction band. A p-(n-) doped semiconductor therefore has an excess of holes (electrons) in the population at a given temperature. These majority carriers dominate the thermoelectric response of the system and determine its sign. Their characteristic energy is somewhat smaller than $\frac{\Delta}{2}$.

With these considerations in mind, let us examine Fig. 6.22, which compares the data reported in three different studies on barely doped silicon. Of the samples studied by Geballe and Hull [1955], one of the lowest n-doped (labelled no. 131 in their table) with an arsenic concentration of $2.75 \times 10^{14}$ cm$^{-3}$. A lightly doped silicon sample with a phosphorus concentration of $2.8 \times 10^{16}$ cm$^{-3}$ was subject of another study, reported

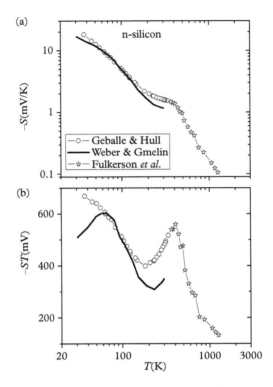

**Figure 6.22** *Seebeck (a) and Peltier (b) coefficients of lightly doped silicon, according to three different studies [Geballe and Hull 1955; Fulkerson et al. 1968; Weber and Gmelin 1991].*

thirty-six years later [Weber and Gmelin 1991]. The third set of data corresponds to a nominally undoped silicon polycrystal studied up to 1300 K [Fulkerson *et al.* 1968]. As seen in the figure, there is a reasonable agreement between the three sets of data, confirming the validity of the extensive set of data [Geballe and Hull 1955] across a wide range of n-doped and p-doped samples.

As seen in the figure, the Seebeck coefficient rapidly decreases with temperature, and its temperature dependence is almost as fast as $T^{-2}$. It is generally believed that the thermoelectric response is mostly dominated by phonon drag, and Herring's theory is mainly elaborated with this set of data in mind [Herring 1954]. Plotting the data as $ST$ versus temperature points to a low-temperature saturation of the Peltier coefficient towards a value close to $\frac{\Delta}{2} = 0.55$ eV. It also reveals an intricate structure in the temperature dependence of the thermoelectric response, a structure which is yet to be understood. There are a number of theoretical computations of the Seebeck coefficients in germanium and silicon [Price 1956; Wang *et al.* 2011]. In these approaches, the experimental data on variation of the room-temperature Seebeck coefficient with carrier concentration has been confronted with calculations and been explained. However, explaining the fine structure of the temperature dependence across the whole range of available data remains beyond these attempts and requires a satisfactory treatment of the phonon-drag contribution.

If silicon is doped further, it will end up becoming a metal. This happens when carriers introduced to the system by doping are numerous enough to detect each other's presence through Coulomb interactions. There is a universal criterion for the critical doping at which this metal–insulator transition occurs [Edwards and Sienko 1978]. In the case of phosphorous-doped silicon, a sharp metal–insulator transition occurs at a critical carrier density of $3.8 \times 10^{18}$ cm$^{-3}$ [Rosenbaum *et al.* 1980]. In a heavily doped semiconductor, on the metallic side of the metal–insulator transition, the Seebeck coefficient displays a temperature-dependence characteristic of a metal. It decreases with temperature and, at low enough temperatures, becomes $T$-linear. This can already be seen in the case of the most doped samples studied by Geballe and Hull (see Fig. 6.20).

Extrinsic silicon, sufficiently doped to be on the metallic side of the metal–insulator transition, has been the subject of two subsequent studies [Weber and Gmelin 1991; Lakner and Löhneysen 1993]. There is a quantitative and impressive agreement between these two sets of data and the free-electron-gas picture. The sample studied by Weber and Gmelin had a concentration of $1.7 \times 10^{19}$ cm$^{-3}$ arsenic atoms (the metal–insulator transition in arsenic-doped silicon occurs at a critical carrier density of $7.8 \times 10^{18}$ cm$^{-3}$ [Newman and Holcomb 1983]). Below 20 K, the Seebeck coefficient was found to be almost $T$-linear, with a slope close to what is theoretically expected. Lakner and Löhneysen studied a phosphorous-doped sample with a concentration of $7.13 \times 10^{19}$ cm$^{-3}$ phosphorous atoms. At this doping level, deep inside the metallic regime, silicon has a well-defined Fermi surface with six anisotropic valleys (Fig. 6.23). Assuming that the carrier concentration in each valley is one-sixth of the total, and the average effective mass to be $m^* = 0.33m_e$, Lakner and Löhneysen calculated the Fermi temperature. The magnitude of this Fermi temperature, $T_F = 682$ K sets the slope of the Seebeck coefficient in the zero-temperature limit as $S = \frac{\pi^2}{3} \frac{k_B}{e} \frac{T}{T_F}$. As seen in Fig. 6.23, the experimental data are is in excellent agreement with this simple picture.

Thus, it is not in alkali or noble metals that the free-electron gas finds its physical realization but, ironically, in a doped semiconductor such as silicon. In Chapter 7, we will see that, in the case of other metallic doped semiconductors, the intimate relation between the low-temperature Seebeck coefficient and the magnitude of the Fermi temperature holds as well.

We finish this discussion with a few words regarding the thermoelectric response at the boundary between lightly doped and heavily doped silicon. According to Löhneysen and collaborators [Lakner and Löhneysen 1993; Liu *et al.* 1996], compensated and uncompensated silicon display qualitatively different behaviours near the metal–insulator transition. The Seebeck coefficient in phosphorous-doped silicon, with a carrier density barely below the critical density, shows a drastic sign change. This sign change is absent in compensated samples with an identical carrier density but containing both phosphorus and boron dopants. This effect has been attributed to strong

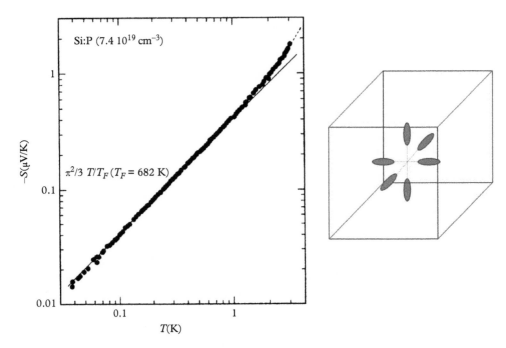

**Figure 6.23** *The Seebeck coefficient of metallic phosphorus-doped silicon (adapted from [Lakner and Löhneysen 1993]). The insert shows the Fermi surface of n-doped silicon.*

electron–electron interactions driving the metal–insulator transition when the system is uncompensated [Löhneysen 2011]. This is another problem missing a quantitative treatment.

The thermoelectric properties of germanium has been studied in less detail than in silicon. In the lightly doped limit [Geballe and Hull 1954], the generic features are quite similar. The gap in germanium is two times smaller than silicon and, unsurprisingly, the magnitude of its Peltier coefficient is smaller. No trace of any investigation of thermo-electric properties in germanium across and beyond the metal–insulator transition has been found.

# 7

# Experimental Survey II: Narrow-Gap Semiconductors

## 7.1 Thermoelectric Materials

This chapter is devoted to three families of narrow-gap semiconductors. They are known as excellent bulk thermoelectric materials, each in a given temperature window: Bismuth telluride ($Bi_2Te_3$) at room temperature, lead telluride (PbTe) above room temperature and bismuth–antimony ($Bi_{1-x}Sb_x$) alloys below room temperature. These materials are among those constituting the subject matter of the main body of research on potentially applicable thermoelectricity, which is concerned with identifying materials which have a large figure of merit, the crucial material-related parameter in a Peltier cooler.

It is easy to see why good thermoelectric materials are expected to be found near the boundary between metals and insulators. As the concentration of mobile electrons is varied, one expects to see opposite trends for electric conductivity and thermoelectric response. With increasing carrier concentration, the electric conductivity rises and the Seebeck coefficient decreases. Therefore, the so-called power factor, defined as $PF = \frac{S^2}{\sigma}$, is expected to peak somewhere in between. With realistic parameters and at room temperature, a peak in power factor is expected near a carrier concentration of $10^{-19}$ cm$^{-3}$ [Mahan, Sales, and Sharp 1997]. This is more than three orders of magnitude lower than the density of mobile electrons in copper ($\sim 8.5 \times 10^{22}$ cm$^{-3}$). There is a third component to the figure of merit: the thermal conductivity, which, at such carrier concentrations, is dominated by the lattice contribution. Good thermoelectric materials principally stand out because of their low lattice thermal conductivity, a property often independent of electronic properties. It is not surprising that most interesting thermoelectric materials contain heavy atoms, which help to damp phonon thermal conductivity.

The three families of narrow-gap semiconductors reviewed in this chapter have been known for many decades as materials with sizeable figures of merit. Intriguingly, all three have attracted new attention during the last few years as topological insulators or topological crystalline insulators [Hasan and Kane 2010]. A topological insulator is a bulk insulator with a particularly robust metallic surface which is due to cooperation between spin–orbit coupling and band inversion. Strong spin–orbit coupling is generally

*Fundamentals of Thermoelectricity*. First Edition. Kamran Behnia.
© Kamran Behnia 2015. Published in 2015 by Oxford University Press.

found in alloys with heavy atoms. Band inversion does not generate large gaps. Heavy atoms and narrow gaps are key ingredients in the physics of topological insulators as well as interesting thermoelectric materials. Hence, an accidental encounter occurs between two conceptually distinct fields of research, focusing on the same materials. However, there is no experimental evidence for a connection between band inversion and enhanced thermoelectric performance.

The recent attention to these materials has led to one important clarification. Almost all the materials dubbed 'topological insulators' fail to qualify as insulators in the real world. However, they often satisfy the most crucial benchmark for metallicity: a sharp Fermi surface. Indeed, most of these narrow-gap semiconductors are sufficiently self-doped to be on the metallic side of the metal–insulator transition. In other words, they are dilute metals. At room temperature, they do not owe their mobile electrons to thermal excitation, but to an unavoidable residual stoichiometry mismatch. The relevant parameter for setting the magnitude of the thermoelectric response in this case is not the size of the semiconducting gap but the magnitude of the Fermi energy, as this magnitude is much lower than that in ordinary metals.

With this in mind, let us ask ourselves a very simple question: what is special about a carrier density of $10^{19} \text{cm}^{-3}$? It happens that we can ask the following question. At what carrier density does the Fermi energy of the free-electron gas becomes equal to room temperature, assuming a free-electron mass ($m^* = m_e$) for carriers ? The equation linking carrier density $n$ and the Fermi temperature $T_F$ is as follows:

$$n = \frac{1}{3\pi^2} \left[ \frac{2m_e k_B T_F}{\hbar^2} \right]^{3/2}$$

(7.1)

Setting $T_F = 300\,\text{K}$ in this equation, one obtains $n = 1.9 \times 10^{19}$ $\text{cm}^{-3}$, which is very close to the optimal carrier density for the maximum power factor in doped semiconductors.

Thus, in the case of the narrow-gap semiconductors discussed in this chapter, one should keep in mind that, while the parent compound is indeed a semiconductor, the doped material, *the one with a large thermoelectric figure of merit*, is not. It is true that, at room temperature, the system is close to the cross over between degenerate and non-degenerate limits. But this should not hide the fact that its ground state is metallic.

An excellent review on the electric, thermal, and thermoelectric properties of all three families discussed in this chapter, as well as other families of thermoelectric materials, can be found in Goldsmid's [2010] book.

## 7.2    Bi$_2$Te$_3$ and Family

Bi$_2$Te$_3$ is the archetype thermoelectric material [Goldsmid 1954]. Discovered in 1954 by H. Julian Goldsmid, this material kept the title of the best thermoelectric material for the next five decades. Even nowadays, commercial Peltier coolers use alloys based on Bi$_2$Te$_3$.

There is a family of $V_2VI_3$ compounds, where $V$ = bismuth/antimony and $VI$ = selenium/tellurium. They all share the same rhombohedral crystal structure. This, as we saw in Chapter 6, is also the crystal structure of elemental bismuth and antimony. However, the resemblance is only qualitative. In the two elements, the rhombohedral angle is close to $57°1'$ [Liu and Allen 1995], which is not very far from the corresponding angle in a cubic lattice (the angle between two body diagonals, which is $60°$). One can think of their elementary cell as a cube pulled along its body diagonal. On the other hand, in both $Bi_2Te_3$ and $Bi_2Se_3$, this angle becomes as small as $24°8'$ [Mishra, Satpathy, and Jepsen 1997]. Little trace of cubic structure has remained here. Instead, the $V_2VI_3$ lattice is clearly layered with successive stacks, which repeat a quintuple building block of $VI(1)–V–VI(2)–V–VI(1)$ [Wiese and Muldawer 1960]. Focusing on the particular case of $Bi_2Te_3$, one can see that there are two distinct sites for the Te atom. Two-thirds of the Te atoms are on site 1, and one third on site 2. When Te is substituted with Se, evidence suggests that the Se atoms prefer to replace the Te atoms on the first site before starting to occupy the second [Wiese and Muldawer 1960].

According to early infrared conductivity measurements [Black *et al.* 1957] confirmed by more recent results [Akrap *et al.* 2014], the gap is 0.15 eV wide in $Bi_2Te_3$ and 0.35 eV in $Bi_2Se_3$. This is in agreement with expectations of the band calculations [Mishra, Satpathy, and Jepsen 1997, Scheidemantel *et al.* 2003, Zhang *et al.* 2009].

There is a more important difference between the two systems. Based on the theoretical band structure, the locus of the band extrema in the Brillouin zone differs between the two. In $Bi_2Se_3$, the situation is simple. The gap is direct and it occurs at the $\Gamma$-point. Therefore, in both n-doped and p-doped $Bi_2Se_3$, the Fermi surface is expected to emerge at the $\Gamma$-point. In $Bi_2Te_3$, on the other hand, the gap is indirect and does not occur at a high-symmetry point of the Brillouin zone. The n-doped and p-doped Fermi surfaces are multi-component, are not located at high-symmetry points, and do not lie along a high-symmetry axis. In both cases, the band structure is largely modified by the spin–orbit coupling. Both the magnitude of the gap and its location are different in the presence or absence of spin–orbit effect [Mishra, Satpathy, and Jepsen 1997; Zhang *et al.* 2009].

Measurements of the Shubnikov-de Haas effect in $Be_2Se_3$ started in the seventies [Köhler and Wöchner 1975]. More recently, in the context of research on topological insulators, this early work was complemented by new studies of quantum oscillations. Thanks to them, the topology of the Fermi surface and its evolution with doping is fairly well established. The Fermi surface emerges as a single, moderately anisotropic ellipsoid located at the centre of the Brillouin zone (See Fig. 7.1). At low doping ($\sim 6.3 \times 10^{17}$ cm$^{-3}$), the effective mass is almost isotropic and about $0.14m_e$ [Butch *et al.* 2010]. As the doping increases, this Fermi surface grows in size and becomes more anisotropic. The Brillouin zone is squeezed along the $z$-axis as a result of the elongated quintuple building block. At a carrier density of the order of $10^{20}$ cm$^{-3}$, therefore, the Fermi surface touches the zone boundary along the $z$-axis and becomes a warped cylinder [Lahoud *et al.* 2013]. This dimensionality crossover is confirmed by the observation of the quantized Hall effect in the bulk system in this range of carrier concentration [Cao *et al.* 2012].

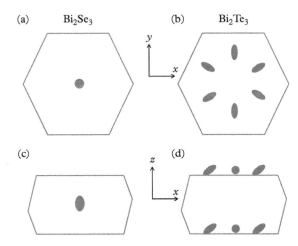

**Figure 7.1** *Fermi surfaces and Brillouin zones in Bi$_2$Se$_3$ (a, c) and in Bi$_2$Te$_3$ (b, d). For the former, the Fermi surface has a single valley and is located at the centre of the Brillouin zone. For the latter, there are six valleys. The long axis of each of the six ellipsoids is tilted by an angle of about 30° off the xy plane.*

As one can expect from its band structure, Bi$_2$Te$_3$ has a more complicated Fermi surface. According to early results by Köhler, the Fermi surface of Bi$_2$Te$_3$ consists of six elongated ellipsoids, each located at a low-symmetry point of the Brillouin zone and tilted by 31° off the horizontal plane of the Brillouin zone (see Fig. 7.2). In both n-doped [Köhler 1976a] and p-doped [Köhler 1976b] cases, the effective mass is of the order of 0.1$m_e$ with an anisotropy of 4 to 6. Its magnitude continuously increases with increasing energy, thus indicating non-parabolic dispersion. Mishra and co-workers have found that the tilt angle and the Fermi-surface topology are both compatible with what is expected from the theoretical band structure [Mishra, Satpathy, and Jepsen 1997]. The evolution of the Fermi surface beyond a doping level of 10$^{19}$ cm$^{-3}$ is yet to be determined.

As-grown crystals in both systems contain vacancies or anti-site substitutions, both of which make them metallic, with a carrier density of about 10$^{19}$ cm$^{-3}$. Using controlled annealing, one can tune the carrier density to between 10$^{17}$ cm$^{-3}$ to 10$^{20}$ cm$^{-3}$ without the addition of extrinsic dopants. While in the case of Bi$_2$Te$_3$ one can obtain both n-doped and p-doped materials, in the case of Bi$_2$Se$_3$, only n-doped samples are produced in this way. Positively charged selenium vacancies are believed to be the source of n-doping in Bi$_2$Se$_3$. Bismuth and tellurium atoms can easily exchange sites and this provides an additional route for doping in Bi$_2$Te$_3$. In the latter system, one can obtain both p-doped and n-doped samples without introducing extrinsic atoms.

A set of recent measurements on solid solutions of Bi$_2$Te$_{2-x}$Se$_x$ by Akrap and co-workers gives an idea of the typical temperature dependence of the Seebeck coefficient in this family [Akrap *et al.* 2014]. As seen in Fig. 7.2, the Seebeck coefficient smoothly

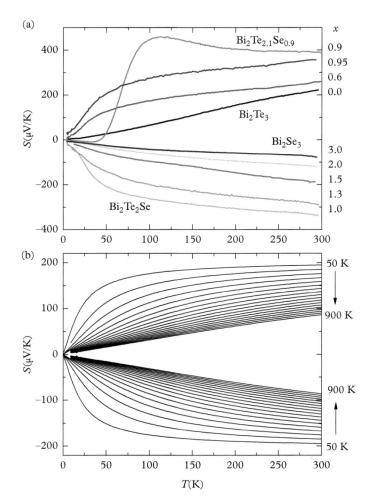

**Figure 7.2** *Evolution of thermopower with selenium substitution in Bi₂ Te₂₋ₓSeₓ. (a) The experimental temperature dependence of the Seebeck coefficient for samples with different x. (b) Theoretical expectation of a simple model for a Fermi-Dirac distribution as the Fermi temperature is varied between 50 K and 900 K (after [Akrap et al. 2014]).*

evolves from positive to negative. For most samples, the evolution can be explained by assuming that the thermoelectric response is purely diffusive and the Fermi temperature smoothly evolves with doping. In other words, there is no striking sign of phonon drag in the signal. Only the sample with a composition that is close to the Bi₂ Te₂ Se composition shows a different temperature dependence. This is the composition at which the gap is largest and apparently the system is no more a degenerate semiconductor. Note that the evolution seen here is *not* directly set by the evolution in the nominal stoichiometry.

Perfect crystals of this family are expected to be just insulators, which is not the case here. Rather, this is a demonstration of the evolution of self-doping with selenium substitution. As-grown crystals are n-type in the case of $Bi_2Se_3$ and p-type in the case of $Bi_2Te_3$, because of an excess of bismuth in the first case and a lack of bismuth in the second.

One may ask at this stage, what is special about $Bi_2Te_3$ to make it the best available thermoelectric material at room temperature? It could not be just the magnitude and temperature dependence of its Seebeck coefficient, which looks very much like what you would expect for any degenerate semiconductor. Then, what is it? Comparing $Bi_2Te_3$ with its sister material, we see that several details stand out. We already saw the first feature, which was the existence and accessibility of both n-type and p-type materials in this particular narrow-gap semiconductor.

The second important feature is the multiplicity of the valleys. Keeping the same carrier density and the same effective mass, one can reduce the Fermi energy in each pocket by a factor of $n^{2/3}$ by dividing the carriers between $n$ pockets of equal size. In that case, provided that the conductivity remains unaffected, the total Seebeck response will be enhanced compared to the single-pocket case.

But the most important feature is the poor phonon thermal conductivity. Because of the anisotropic lattice structure, the phonon thermal conductivity is two times lower out-of-plane than in-plane. The extremely low lattice thermal conductivity is probably the key factor for pushing $Bi_2Te_3$ to the top place in the figure-of-merit league. After all, in a doped semiconductor at room temperature, when the resistivity is about 1 m$\Omega$ cm, a Seebeck coefficient of 200 $\mu V/K$ does not sound exceptional, but a lattice thermal conductivity as low as 1.6 W/mK does. Thus, the exceptionality of $Bi_2Te_3$ owes more to its particular crystal structure than to its electronic properties. Phonon heat conduction in $Bi_2Te_3$ is one of the lowest known among common insulators [Keyes 1959]. As we will see below, when comparing it with PbTe, the phonon mean-free-path in this system at room temperature becomes a few angstroms, much shorter than the height of the elementary cell and barely longer than the interatomic distance.

## 7.3   PbTe and Other IV–VI Salts

PbTe has been around as a prominent thermoelectric material since the 1950s. As early as 1959, a five-watt radioisotope thermoelectric generator was made in the United Sates using the Seebeck effect in PbTe to convert radioisotope heat to electricity (for recent reviews and historical accounts, see [Pei *et al.* 2011a] and [Lalonde *et al.* 2011]). Meanwhile, in the former Soviet Union, Ioffe was leading a school of research exploring interesting semiconductors for thermoelectricity, including this particular narrow-gap one. During recent years, PbTe has emerged again as a bulk material with a very competitive figure of merit at temperatures above room temperature (see Fig. 7.3).

PbTe belongs to a family of IV–VI compounds, which include lead setenied (PbSe), lead (II) sulphide (PbS), tin telluride (SnTe), germanium telluride (GeTe), and others. These are all narrow-gap semiconductors with rock-salt structure and close to a ferro-electric instability. Besides thermoelectricity, the small gap (0.18 eV in PbTe) has made

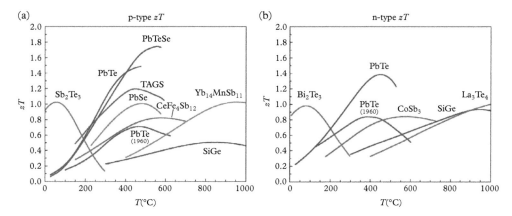

**Figure 7.3** *Thermoelectric figure of merit for various n-type and p-type materials (after [Lalonde et al. 2011]).*

these materials interesting for applications in infrared devices. More recently, new attention has focused on band inversion in solid lead–tin–tulluride ($Pb_xSn_{1-x}Te$) alloys. In SnTe and PbTe, a direct band gap of comparable magnitude is located at the L-point of the Brilluoin zone. However, a particularity of this gap was noticed a long time ago [Dimmock, Melngailis, and Strauss 1966]. The conduction and valence bands are inverted between the two compounds. Therefore, as one goes from PbTe to PbSe, a Dirac point emerges close in the band structure of $Pb_xSn_{1-x}Te$ near $x = 0.3$. The inversion occurs because of the difference in the magnitude of the relativistic effects of the heavier lead compared to the lighter tin.

This family of materials is very close to having ferroelectric instability. Their dielectric constant is three orders of magnitude larger than vacuum. As a result of the small gap, the effective mass is small. The combination of a high dielectric constant and a small gap makes the Bohr radius very long. The metal–insulator transition is therefore expected to occur at very small doping concentrations. Indeed, early studies found that nominally intrinsic PbTe is a dilute metal with carriers of surprisingly high mobility at low temperatures. The 4.2 K Hall mobility was found to become as large as $8 \times 10^5$ cm$^2$ V$^{-1}$ s$^{-1}$ at a carrier density of $10^{18}$ cm$^{-3}$ [Allgaier and Scanlon 1958].

Careful Shubnikov-de Haas measurements in 1970 showed that, in this range of carrier density ($\sim 3 \times 10^{18}$ cm$^{-3}$), the Fermi surface of p-doped PbTe consists of four ellipsoids oriented along the $< 111 >$ orientation [Burke, Houston, and Savage 1970]. This is in agreement with band-structure calculations, which predict a gap at the L-point of the Brillouin zone. The anisotropy of the effective mass was found to be about 13, with the lighter mass becoming as small as $0.036m_e$. The evolution of this Fermi surface with increasing doping was the subject of a later Shubnikov-de Haas study [Jensen, Houston, and Burke 1978], which found that, save for a two-fold increase in the effective mass, this description of the Fermi surface consisting of four identical ellipsoids remains valid up to a carrier density of $4.5 \times 10^{19}$ cm$^{-3}$. Even the anisotropy of each ellipsoid barely changed in this range of concentration. However, at this concentration, the model ceased to work, presumably as a consequence of a change in Fermi-surface topology. The low-temperature mobility had become so small by this carrier concentration that the detection of quantum oscillations would need magnetic fields larger than what was used at that time. Such studies are required in order to find a solid answer to the following question: what happens to the Fermi surface as the Fermi energy exceeds 20 meV and the carrier concentration becomes larger than $0.03h^+$ per formula unit?

A possible answer to this question was provided by a study of the magnetoresistance of warm holes [Sitter, Lischka, and Heinrich 1977]. In this experiment, the angle-dependent magnetoresistance of p-doped PbTe samples with a carrier density of $2-3 \times 10^{18}$ cm$^{-3}$ was measured at high temperatures, and an inversion of angular extrema was detected between 359 K and 390 K. The authors interpreted their result as the occupation by thermally excited carriers of a new component of Fermi surface. A quantitative analysis suggested that, compared to the lower band, this new upper component had a similar mass anisotropy with an inverted sign. Therefore, the authors suggested that the new components are eight ellipsoids at the $\Sigma$-point of the Brillouin zone (See figure 7.4a).

Band-structure calculations predict that, at a much higher carrier concentrations, a unique connected Fermi surface of connected 'pipes' [Parker, Chen, and Singh 2013] emerges (see Fig. 7.4c). The intermediate topology indicated by the warm angle-dependent magnetoresistance studies does not appear incompatible in these calculations. With increasing doping, the four L ellipsoids and the eight $\Sigma$ ellipsoids grow in size and could interconnect to give rise to the 'pipes'. A similar evolution is seen in a set of band calculations in the case of the sister compound SnTe [Littlewood *et al.* 2010].

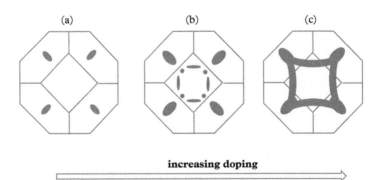

increasing doping

**Figure 7.4** *Evolution of the Fermi surface in p-doped PbTe. (a) The Fermi surface determined by experimental measurements of Shubnikov-de-Haas effect consists of 4 L ellipsoids. (b) According to warm magnetoresistivity measurements[Sitter, Lischka, and Heinrich 1977] 8 $\Sigma$ ellipsoids emerge at higher doping. These pockets have not been seen by quantum oscillations. (c) According to band calculations [Parker, Chen, and Singh 2013], there is a single Fermi surface when the Fermi energy is as large as 0.23 eV. This is yet to be confirmed by experiment.*

Very large figures of merit, approaching 1.5 at 800 K, have been reported in p-type PbTe doped with thallium [Heremans *et al.* 2008], and with sodium [Pei *et al.* 2011a]. An even larger $ZT$ has been reported in $PbTe_{1-x}Se_x$ doped with sodium [Pei *et al.* 2011b]. In all these cases, the doping level is about 0.005 to 0.02 holes per formula unit. As we saw above, the precise topology of the Fermi surface at this doping level has not yet been established by measurements of quantum oscillations. It is still unknown if the Seebeck coefficient is the response of twelve independent L and $\Sigma$ valleys [Pei *et al.* 2011b] or a single interconnected entity with intricate pipe-like topology [Parker, Chen, and Singh 2013].

Another open question is the variation in the magnitude of the Seebeck coefficient with different dopants. At a carrier density of a few $10^{19}$ cm$^{-3}$, the room-temperature Seebeck coefficient is larger in thalium-doped crystals than in self-doped or sodium-doped crystals. Interestingly, this is the carrier density at which superconductivity sets in, with a critical temperature significantly higher in thalium-doped samples than in the others [Matsushita *et al.* 2005]. It has been suggested that thalium-doping leads to a resonant density of states, and this is both the origin of the enhanced Seebeck coefficient in thalium-doped PbTe [Heremans 2008] and superconductivity. A study of the quantum oscillations of PbTe in this range of carrier concentration, pinning down the Fermi surface topology for different dopants, would shed more light on this.

In summary, doped PbTe is a dilute metal with a tantalizing Fermi surface, which has yet to be sorted out in detail. Meanwhile, one can try to put the appeal of PbTe for thermoelectricity at high temperatures in a few rough words. PbTe replaces $Bi_2Te_3$ above room temperature because it remains a degenerate system fermion at higher temperatures, thanks to the lower effective mass of its carriers. This allows the system to have a higher Fermi energy at the same carrier concentration.

## 7.4   $Bi_{1-x}Sb_x$ Alloys

As elements, bismuth and antimony, as well as their Seebeck coefficients, were briefly reviewed in Chapter 6. They are semimetals with a multi-valley Fermi surface, composed of hole-like and electron-like components. They share the same crystal structure and a very similar Brillouin zone [Issi 1979]. The components of the Fermi surface, however, are located at different points and with a sizeable difference in volume. The carrier concentration in antimony is larger by two orders of magnitude. One may expect therefore that, by alloying bismuth and antimony, one would have the opportunity to witness a Lifshitz transition, a thermodynamic phase transition associated with a change in the topology of the Fermi surface [Lifshitz 1960]. What happens in the real alloy, however, is more complicated. Between the two semi metals lies a region where a gap opens up between conduction and valence bands, and the system becomes a narrow-gap semiconductor. This was found by an early study of transport properties in $Bi_{1-x}Sb_x$ alloys [Jain 1959]. An activated resistivity was found for $0.06 < x < 0.4$, with a small gap peaking to 14 meV at $x = 0.12$.

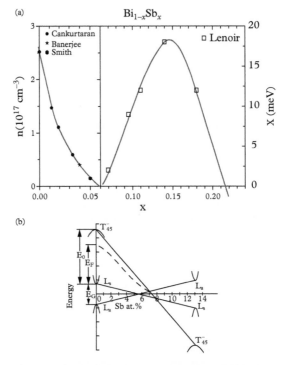

**Figure 7.5** *Semi metal to semiconductor transition in Bi$_{1-x}$Sb$_x$. (a) Evolution of carrier concentration with antimony doping extracted from the frequency of quantum oscillations [Smith 1962; Cankurtaran, Celik, and Alper 1985; Banerjee et al. 2008] and the activation gap resolved by resistivity [Lenoir et al. 1996]. (b) Evolution of band structure, energy gaps, and the Fermi level of bismuth with antimony alloying. Reprinted with permission from [Hiruma and Miura 1983]. Copyrighted by the Physical Society of Japan.*

The establishment of the band structure and the Fermi-surface topology in bismuth during the following decades led to a quantitative understanding of the evolution of electronic properties with antimony doping. The schematic evolution [Hiruma and Miura 1983] is sketched in Fig. 7.5. Three distinct energy scales are modified by antimony doping. First of all, the small gap at the L-point, $E_G$ ~15 meV, decreases very quickly and vanishes between $x = 0.04$ and $x = 0.06$. At this critical doping level, the energy dispersion of carriers at the L-point becomes linear, and they become Dirac electrons. Beyond this doping level, a gap opens again at the L-point. The two L bands (L$_a$ and L$_s$) are now inverted. The second energy scale is the negative (overlap) gap ($E_0$ ~38 meV), the distance between the bottom of the conduction band at the L-point and the top of the valence band at the T-point. It vanishes at a critical doping close to $x = 0.07$.

Beyond this doping level, the system is no more a semimetal but a semiconductor. The third relevant energy scale is the Fermi energy. It shifts in order to keep charge neutrality as long as there are metallicity and mobile carriers. The shift ensures that the electron-like and hole-like components of the Fermi surface, while different in topology, remain equal in volume.

The inversion of bands in the semiconducting regime of the $Bi_{1-x}Sb_x$ alloy made it the first identified topological insulator [Fu and Kane 2007]. However, like the two other materials reviewed in this chapter, the bulk material failed to be a true insulator, presumably because of uncontrolled and unavoidable doping. The situation is a bit different here, though. Contrary to the two previous cases, if bismuth and antimony atoms exchange their sites, no doping consequences would arise. However, an infinitesimal amount of vacancies or extrinsic impurities would shift the Fermi energy. The Bohr radius is expected to be long, because of the combination of a light electron mass ($m^* \simeq 0.001 - 0.2m_e$) and a sizeable dielectric constant ($\epsilon \sim 100\epsilon_0$). As a consequence, a very small amount of doping is sufficient to put the system on the metallic side of the metal–insulator transition.

The evolution of the Fermi surface below the critical doping for the semi metal-to-semiconductor transition has been explored using quantum oscillations by Brandt and collaborators in the former Soviet Union in the late sixties and early seventies [Brandt, Lyubutina, and Kryukova 1968; Brandt and Chudinov 1971]. Experiments documented the shrinking of the hole and electron pockets with antimony doping. During that period, much attention was paid to the possible irruption of an excitonic insulator near the critical doping at which the Fermi surface vanishes. Later studies of quantum oscillations, using ultrasonic attenuation [Cankurtaran, Celik, and Alper 1985] or the Nernst coefficient [Banerjee *et al.* 2008] as experimental probes, confirmed the early results and indicated that the Fermi surface would vanish around $x \sim 0.07$. In this doping range, this becomes the smallest bulk Fermi surface ever detected by experiments in condensed-matter physics. The possible existence of a Fermi surface beyond $x = 0.07$ as a consequence of extrinsic doping remains an open question.

As early as 1962, Smith and Wolfe found that $Bi_{1-x}Sb_x$ alloys (with $x \sim 0.12$) are interesting thermoelectric materials at cryogenic temperatures, and their thermoelectric performance is enhanced by the application of a small magnetic field [Smith and Wolfe 1962; Wolfe and Smith 1962]. Both these features were confirmed ten years later [Yim and Amith 1972]. Two more decades later, Lenoir and co-workers carried out an extensive study of transport coefficients in carefully prepared samples of $Bi_{1-x}Sb_x$ with homogeneous antimony concentrations [Lenoir *et al.* 1996]. This remains the most detailed information on the resistivity, Seebeck coefficients, and thermal conductivity of these alloys available.

Elemental bismuth has the largest thermoelectric figure of merit among elements. At room temperature, when heat flows along the trigonal axis, $ZT$ is as large as 0.3. As we saw in Chapter 6, the Seebeck coefficient is negative and as large as 100 $\mu VK^{-1}$ along the trigonal axis. The negative sign implies a domination of the thermoelectric response by electron-like carriers, which are more mobile than hole-like carriers. The lattice thermal conductivity is remarkably low at room temperature (1–2 $WK^{-1} m^{-1}$), giving rise to the sizeable figure of merit. As bismuth is cooled down, however, while the Seebeck coefficient remains more or less flat, the enhancement of thermal conductivity (due to an eightfold increase in lattice conduction between room temperature and 100 K) is not compensated by an increase in electric conductivity. Therefore, $ZT$ decreases with decreasing temperature.

We can see what makes bismuth–antimony alloys interesting thermoelectric materials around the liquid nitrogen temperature by comparing them with bismuth (see Fig. 7.6). At room temperature, these alloys look very much like bismuth regarding their electric, thermoelectric, and thermal transport. At first, this may look surprising. But the gap at the L-point is so small that it does not matter much at room temperature. Most carriers are thermally excited. The difference between pure bismuth and the alloys emerges as the system is cooled down and with the depopulation of thermally excited carriers in the conduction band. This leads to an enhancement of both the Seebeck coefficient and electric resistivity with cooling in the alloys. The power factor slightly increases compared to pure bismuth. However, the most important difference regards the lattice thermal conductivity. In contrast to bismuth, it remains flat in the alloys between room temperature and 100 K. Phonons cannot travel very far because, at 7 per cent concentration, the distance between two antimony atoms is just two to three times the interatomic distance, and this sets the phonon mean-free path between 100 K and 300 K. In contrast, in bismuth, as the temperature decreases, the phonons can travel further and further as a result of the decrease in the phonon–phonon scattering.

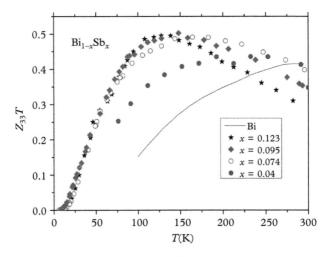

**Figure 7.6** *Figure of merit along the trigonal axis, $Z_{33}T$, in bismuth–antimony alloys (adapted from [Lenoir 1996]). It presents a peak, absent in pure bismuth. This is mostly thanks to a temperature-independent lattice conductivity over a wide temperature range.*

In both bismuth and bismuth–antimony alloys, the Seebeck coefficient is modified by the application of a small magnetic field. This has a fortunate consequences for thermo-electric performance. Indeed, this magneto-Seebeck effect is negative and thus, the absolute value of the thermoelectric response enhances with increasing field. As Wolfe and Smith [1962] first noticed and Yim and Amith [1972] confirmed afterwards, the figure of merit increases at low fields and then peaks at a magnetic field of about 1 T. Beyond this field range, the rise in magnetoresistance is not compensated by the rise in the Seebeck coefficient, and the figure of merit decreases (see Fig. 7.7). Since a magnetic field of this order of magnitude can be generated by permanent magnets, this provides another avenue for application.

Since carriers in bismuth and bismuth–antimony alloys are highly mobile, the Nernst coefficient is very large. In the presence of a magnetic field of 1 T, the transverse electric field generated by the thermal gradient is not negligible compared to the lon-gitudinal one. In contrast to a Peltier cooler, an Ettingshausen cooler would use the Nernst effect and not the Seebeck effect. One advantage of an Ettingshausen cooler is

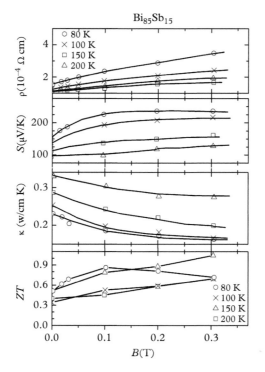

**Figure 7.7** *The variation of the three transport coefficients and the figure of merit with magnetic field in $Bi_{0.85}Sb_{0.15}$ (adapted from [Yim and Amith 1972]).*

the possibility of making an infinite-stage cooler based on an optimal sample geometry [Kooi *et al.* 1968]. This strategy for cooling was employed in a spectacular 1964 experiment [Harman *et al.* 1964]. A bismuth single crystal was made in the presence of a magnetic field of 11 T to reach 196 K from room temperature. This 100 K cooling was achieved by injecting 60 A in the centimetre-sized crystal. This is the largest temperature difference ever reported to be attained by a single-stage thermoelectric cooler.

This experiment used a magnetic field of 11 T, which was exceptionally large at that time, but widely available these days. Nevertheless, and somewhat surprisingly, there is no report on any attempt to reproduce the results, let alone to surpass it. On the other hand, an Ettingshausen cooler, based on a permanent magnet and a bismuth–antimony crystal, has been made which confirms the validity of this approach. The crystal was shaped into the optimal geometry of an infinite-stage cooler. When its hot end was kept at 160 K, a 40 K temperature difference was generated in the presence of 0.75 T [Scholz *et al.* 1994]. In this temperature range, this performance is superior to what can be obtained with $Bi_2Te_3$ at zero magnetic field.

Jandl and Bikholz have shown that doping $Bi_{0.95}Sb_{0.05}$ with small amount of tin (in the range of 70 to 440 ppm) leads to a significant enhancement of the thermomagnetic figure of merit [Jandl and Birkholz 1994]. According to what we know about the evolution of band structure in bismuth with tin doping [Kilic and Celik 1994], this concentration of acceptors is sufficient to make the system a dilute metal. The Fermi surface is a single hole-like Fermi surface at the T-point at small p-doping and becomes a multi-valley Fermi surface with several hole pockets at higher concentrations. But this has not been confirmed by any direct study of quantum oscillations in tin-doped $Bi_{0.95}Sb_{0.05}$. This largely unexplored field of investigation may have a promising future, thanks to the increasing possibility to employ permanent magnets at cryogenic temperatures for improving thermoelectric performance.

## 7.5    Phonon Mean-Free-Path and Interatomic Distance

What qualifies these materials for applied thermoelectricity is their low lattice thermal conductivity. Lattice conductivity in $Bi_2Te_3$ is 1.6 $WK^{-1}m^{-1}$ at room temperature and decreases by a factor of two through doping with $Sb_2Te_3$ [Goldsmid 2010]. In PbTe, it is about 2 $WK^{-1}m^{-1}$ at room temperature and decreases with increasing temperature to attain a quasi-constant plateau twice lower in magnitude around 600 K. This should correspond to the intrinsic limit set by atomic distances. As a matter of fact, the recently revised magnitude of the figure of merit in PbTe at 700 K (1.2 instead of 0.7) owes more to careful measurement of thermal conductivity, indicating a thermal conductivity almost two times lower than what was believed before, than any particular improvement in material design [Pei *et al.* 2011a].

The phonon mean-free-path can be estimated by the kinetic formula according to which $\kappa_{ph} = 1/3 C_{ph} v_s \ell_{ph}$. The Debye temperature in both systems is well below room

**Table** 7.1 *Lattice thermal conductivity in the two thermoelectric materials indicates an extraordinary short phonon mean-free-path*

| | $Bi_2Te_3$ | $PbTe$ |
|---|---|---|
| Debye temperature $[\Theta_D]$(K) | 155 | 150 |
| $\kappa_L$(WK$^{-1}$ m$^{-1}$) | 1.4 (300 K) | 0.8 (750 K) |
| $C_v$(JK$^{-1}$ cm$^{-3}$) | 1.56 | 1.22 |
| $v_s$(km s$^{-1}$) | 2.1 | 3.6 |
| $\ell_{ph} = \frac{3\kappa}{C_v v_s}$ (nm) | 1.28 | 0.54 |
| lattice parameter (nm) | 3.05 (*c*-axis) | 0.65 |
| shortest interatomic distance [*d*] (nm) | 0.32 | 0.32 |
| $\frac{k_B^2}{\hbar}\frac{\Theta_D}{d}$ (WK$^{-1}$ m$^{-1}$) | 0.82 | 0.84 |

temperature and, therefore, the lattice specific heat at room temperature and beyond becomes independent of temperature. Its magnitude in this temperature range is set by the universal Petit-Dulong value. This is confirmed experimentally [Parkinson and Quarrington 1954]. The sound travels slowly in both systems. As seen in Table 7.1, the estimated phonon mean-free-path is shorter than the lattice parameter in both systems. It approaches the shortest atomic distance, which sets the lowest phonon wavelength and the shortest conceivable length scale in the system.

In contrast to the rhombohedral $Bi_2Te_3$, there is nothing exotic about the cubic crystal structure of PbTe. Such a short phonon mean-free-path is therefore intriguing. The low thermal conductivity in PbTe has attracted much recent attention. Neutron diffraction points to a strong anharmonic coupling with ferroelectric transverse optic mode, which drastically reduces the mean-free-path of heat-carrying longitudinal acoustic phonons [Delaire *et al.* 2011]. If this is indeed the case, then the proximity to ferroelectric instability may play a major role, with interesting consequences for the search of new materials.

Above Debye temperature, all available phonon modes are excited and the specific heat becomes temperature independent. The strong interaction between phonons confine them to a space as small as a lattice parameter. Being a solid in this context of strongly interacting phonons seems to be the single most important factor to qualify as a good thermoelectric material. The quest for such materials has been a subject of recent attention [Morelli, Jovovic, and Heremans 2008].

The sound velocity $v_s$ is linked to the Debye temperature $\Theta_D$ through the interatomic distance $d$. Namely, one can write $v_s \simeq \frac{d\Theta_D}{\hbar}$. The Petit-Dulong law implies that the specific heat per volume is $C_p \simeq \frac{3k_B}{d^3}$. The lattice conductivity is lowest when the phonon mean-free-path becomes as short as the interatomic distance. Thus,

$$\kappa_L^{min} = \frac{k_B^2}{\hbar}\frac{\Theta_D}{d} \tag{7.2}$$

The minimum lattice conductivity according to this expression for $Bi_2Te_3$ are given in Table 7.1. One can see that the lattice conductivity measured in PbTe is very close to this minimum.

# 8

# Experimental Survey III: Correlated Metals

The Coulomb interaction between electrons is neglected in the band picture of solids. However, there are cases in which this interaction is believed to play a major role. These are the subject matter of research on 'strongly correlated electrons'. This chapter is devoted to a review of experiments probing the thermoelectric response in such systems. This is an extensive research activity and very hard to describe in such a reduced space. The research has been going along many different directions and in many cases, to be honest, it is still hard to make sense of the data. This should only mildly surprise the reader of the previous chapters of this book, given the situation of elemental systems described in Chapter 6. However, according to one condensed-matter folklore, widely spread in the beginning of the twenty-first century, Fermi liquids and their transport properties are well understood. It has been widely forgotten that, even in the case of Fermi liquids, the agreement between experimental thermoelectricity and its theoretical understanding lags far behind what has been achieved in the case of electric or thermal conductivity.

The following sections of this chapter will each focus on one family of such correlated metals. In each case, we will try to make sense of experiments studying thermoelectric response in the context of the principal questions raised around the paradigms mobilizing the scientific community concerned by a research theme. These experimental sections are preceded by a discussion of the Heikes formula and its origins. This formula, as the expected thermoelectric response in a Hubbard model, has heavily influenced the interpretation of the experimental data obtained on numerous doped Mott-Hubbard insulators.

## 8.1 Hopping Electrons and the Heikes Formula

Heikes introduced his formula in Chapter 4 of his 1961 book entitled *Thermoelectricity: Science and Engineering* [Heikes and Ure 1961]. He considered a (narrow-band) semiconductor in which the carriers are not subject to a Fermi-Dirac distribution, but can

*Fundamentals of Thermoelectricity*. First Edition. Kamran Behnia.
© Kamran Behnia 2015. Published in 2015 by Oxford University Press.

hop from one site to the other. Defining the carrier concentration to be $c$, he argued that the Seebeck coefficient should include a component expressed as

$$S_H = \frac{k_B}{e} ln \left( \frac{c}{1-c} \right) \tag{8.1}$$

Besides stating that it represented 'entropy of mixing', he did not elaborate much about the origin of this term. In classical thermodynamics, this refers to the increment in entropy caused by mixing two previously separated gases. Consider two distinct ideal gases at the same temperature and pressure separated by a dividing wall. After removing the wall, the system will have an entropy larger than the sum of the entropies of its initial subsets. In the context of hopping electrons, the two subsets are occupied and unoccupied sites. Their density is, respectively, $c$ and $1 - c$. Electron hopping mixes the two subsets.

Heikes formula appeared again in an article Austin and Mott [1969] wrote a few years later that discusses the case of the thermoelectric response of a partially compensated semiconductor. They argued that if polarons are localized, in the absence of any energy distribution among hopping sites, then the expected thermoelectric response would match the Heikes formula.

*Thermopower in the Correlated Hopping Regime* was the title of an influential paper by Chaikin and Beni [1976] giving an elegant derivation of the Heikes formula and its extensions in the presence of electron interactions. The authors began by considering spinless fermions, which, because of the Pauli exclusion principle, cannot simultaneously occupy the same site. This is the simplest approximation of the Hubbard model. If there are $N$ particles trying to occupy $N_A$ available sites, the number of possible configurations would be

$$g = \frac{N_A!}{N!(N_A - N)!} \tag{8.2}$$

Using Stirling's approximation $(ln(n!) \simeq nln(n) - n)$, one finds that

$$ln(g) \simeq ln \left( \frac{1-c}{c} \right) \tag{8.3}$$

where $c = N/N_A$ is the ratio of particle number to site number, in other words, the carrier concentration. This clarifies the origin of Eq. 8.1. The 'entropy of mixing' invoked by Heikes is the entropy associated with the coexistence of two species of sites, occupied and unoccupied. But, how about real electrons, which are fermions with spins? Chaikin and Beni argued that, in this case, Eq. 8.2 becomes

$$g = \sum_{N_\uparrow=0}^{N} \frac{N_A!}{N_\uparrow!(N_A - N_\uparrow)!} \frac{N_A!}{N_\downarrow!(N_A - N_\downarrow)!} \tag{8.4}$$

Here, $N_\uparrow$ and $N_\downarrow$ represent the number of spin-up and spin-down electrons, and thus $N = N_\uparrow + N_\uparrow$. In spite of its apparent complexity, Stirling's approximation leads to an expression slightly different from the previous case:

$$ln(g) \simeq ln\left(\frac{2-c}{c}\right) \tag{8.5}$$

Note that, in this last case, it is assumed that two electrons of opposite spin and which do not have the same quantum state, can occupy the same site. Now, if the on-site repulsion is strong, two electrons, either of the same or opposite spin, cannot occupy the same site. In that case, one has

$$g = \frac{2^N N_A!}{N!(N_A - N)!} \tag{8.6}$$

which this leads to

$$ln(g) \simeq ln(2)\left(\frac{1-c}{c}\right) \tag{8.7}$$

In this approximation, which is basically equivalent to a simple Hubbard model, the expression for the Seebeck coefficient would be

$$S_H = \frac{k_B}{e} ln\left(\frac{2c}{1-c}\right) \tag{8.8}$$

Chaikin and Beni considered other cases too. In one of them, there is a repulsive interaction between nearest neighbours. In another, the on-site interaction is positive. In each case, a slightly different expression for thermoelectric response was found [Chaikin and Beni 1976].

Now, consider Eq. 8.8, which represents the expected thermoelectric response of a Hubbard system when the on-site repulsion is much larger than temperature, which is in turn much larger than the Fermi energy. This is a strongly interacting and non-degenerate system of electrons with spins. When $c = 1/2$, the expected Seebeck coefficient is independent of temperature and equal to $\frac{k_B}{e} ln(2)$. Beni, Chaikin, and Kwak argued that this represents the $k_B ln(2)$ spin entropy of charged carriers and proposed that it accurately describes the experimentally measured thermoelectric power in tetracyanoquinodimethane (TCNQ) salts [Beni, Kwak, and Chaikin 1975].

The relevance of Eqs 8.1 and 8.8 to experimental data remains an open question. The entropy that degenerate electrons can carry with them is expected to decrease with cooling. When electrons are not degenerate, on the other hand, one would expect that as they are cooled down, the entropy they carry would increase with decreasing temperature. These two limits represent the metallic and non-metallic regimes of thermoelectricity. In this non-interacting picture, there can always be a regime of quasi-flat Seebeck coefficients along the road from metallic to insulating thermoelectricity.

It is often hard to distinguish between this and the relevance of the Hubbard-type, spin-related, temperature-independent thermoelectricity.

Note also that according to both equations, one should expect very large values of Seebeck coefficient when $c$ is small. Of course, this happens because when a few carriers find themselves in a desert of empty sides, there are many possible configurations. But, can each electron carry this huge amount of entropy with it?

## 8.2    Organic Conductors

During the 1970s, the synthetic metal, tetrathiofulvalene-TCNQ, often abbreviated to TTF-TCNQ, attracted much attention. The molecule of this salt consists of a donor (TTF) and an acceptor (TCNQ). The solid, made of alternating chains of TCNQ anions and TTF cations, is a one-dimensional conductor. During the following decades, a variety of such organic conductors were discovered and studied. Most of them are very anisotropic conductors in which electrons can travel easily along one or two directions and not along the other(s). The most famous stars in this galaxy are Bechgaard salts, based on tetramethyl-tetraselena-fulvalene (TMTSF), Fabre salts, based on its sulphur analogue, tetramethyl-tetrathia-fulvalene (TMTTF), and the ET-family based on bis-ethylenedithio-tetrathiafulvalene (BEDT-TTF).

In spite of their chemical complexity, when the ground state of these systems is metallic, the Fermi surface becomes one of impressive simplicity; slightly warped sheets or quasi-perfect cylinders. As a consequence, they have proved to be a neat laboratory of concepts in fundamental theory of electrons in solids. In the words of Paul Chaikin, all kinds of electron transport known to mankind (metallicity, semiconductivity, superconductivity, quantum Hall effect, charge-density-wave gliding) can occur in a Bechgaard salt, just by tuning temperature, magnetic field, and pressure. There are numerous reviews of these systems. We refer the reader to one of the most recent ones by the founding father of organic superconductivity [Jérome 2012]. Here, we review a number of studies on the thermoelectric response of organic conductors.

The first study of the thermopower of TTF-TCNQ [Chaikin *et al.* 1973] found a reproducible (i.e. sample-independent) negative $T$-linear Seebeck coefficient could be measured that at temperatures exceeding 100 K. The slope of the linear temperature dependence ($\sim 0.1 \ \mu V \ K^{-2}$) was in agreement with the estimated magnitude of the Fermi energy. It displayed a deviation from this linear behaviour at lower temperatures, before abruptly changing sign at a metal–insulator phase transition at 56 K. The interpretation was conventional. At high temperatures, it was a one-dimensional metal with a linear Seebeck coefficient, and this quasi-one-dimensional conductor was fragile in front of a metal–insulator transition.

Experiments on other organic conductors, however, led to more surprising results and more controversial interpretations. It was found that, in a number of salts with one electron per two TCNQ molecules, the Seebeck coefficient saturates to a constant value of about −60 $\mu V/K$ (see Fig. 8.1). This was first pointed out by Buravov *et al.* [1971], who suggested that the spin degrees of freedom were quenched and that thermopower

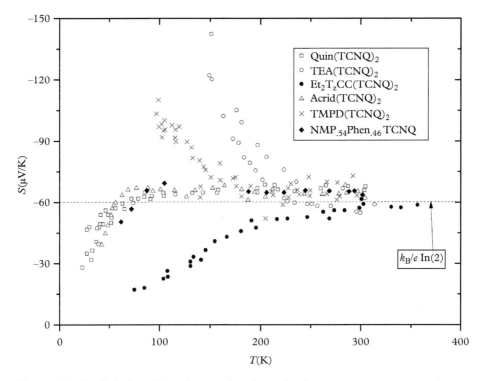

**Figure 8.1** *The Seebeck coefficient in a number of organic salts saturates to a constant value close to 60 μV/K in a number of organic salts (adapted from [Chaikin et al. 1979]).*

was due to the motion of localized carriers, moving from site to site via thermodynamically activated jumps. A diametrically opposite interpretation was proposed by Beni, Kwak, and Chaikin [1975] who suggested that the high-temperature thermopower in these systems was caused by spin entropy. They argued that the systems under study were described by the Hubbard model in the strong-coupling limit. The expected thermopower in this case according to Eq. 8.8 for $c = 1/2$ is $\frac{-k_B}{e} ln(2) = -59.8\ \mu$V/K, close to what was experimentally found.

A further argument in favour of this latter scenario was provided by Chaikin, Kwak, and Epstein [1979], who measured the magneto-Seebeck effect in one of these organic salts, namely quin(TCNQ)$_2$ and found a field-induced reduction of thermopower in the temperature-independent region. The magnitude of the change was small. A magnetic field of 18.5 T reduced the magnitude of the Seebeck coefficient by 1.5 per cent at T = 78 K. However, the effect was five times larger than the measured magnetoresistance and isotropic. Moreover, the magnitude was in fair agreement with the expected reduction in the entropy of free spins caused by the application of the magnetic field. Chaikin and co-workers concluded that the system is in the strong-coupling Hubbard limit with spin and charge degrees of freedom completely separated.

What was not clear in this scenario was the fate of charge entropy. After all, these are conductors with a finite conductivity, pointing to an orbital degree of freedom for electrons. How can one neglect its contribution to the thermoelectric response? To tackle this issue, Chaikin and collaborators invoked a perfect electron-hole symmetry, which would totally cancel the orbital contribution to the Seebeck coefficient [Chaikin, Kwak, and Epstein 1979]. Since such a perfect asymmetry does not arise in any other system that we know, one can hardly consider the argument convincing.

An elegant solution to this problem was suggested by Conwell, who argued against the applicability of an extended Hubbard model with an infinite on-site repulsion, a large nearest-neighbour repulsion, and negligible hopping to the systems under study. Instead, he proposed a model in which a temperature-independent Seebeck coefficient of expected magnitude could arise from charged carriers [Conwell 1978]. In this model, the upper and lower Hubbard bands sandwich the chemical potential and the Seebeck coefficient has an electron-like and a hole-like component. Each of these two components would have a temperature-dependent ($\propto 1/T$) contribution to thermoelectric response, which would cancel out in the case of perfect symmetry. Conwell argued that in the case of strong on-site repulsion forbidding two electrons of either spin orientation to occupy the same site, the chemical potential would shift as a function of temperature to keep the charge neutrality. In this case, if the direct contribution of the lower and upper Hubbard bands to the Seebeck coefficient cancel each other, a residual contribution with a magnitude of $-\frac{k_B}{e}ln(2)$ survives. Both the magnitude of the Seebeck coefficient and its lack of temperature variation in the 2:1 TCNQ salts could find an explanation in this picture.

In the following decades, the thermoelectric response in other quasi-one-dimensional organic conductors was put under scrutiny. These were the Fabre salts, with a generic formula of $(TMTTF)_2X$, where X is an anion, which can be an elemental ion such a bromine or an assembly of atoms such as $BF_4$, $NO_3$, or $PF_6$. If sulfur is replaced by selenium, one obtains the Bechgaard salts with the generic formula $(TMTTF)_2X$. In Bechgaard salts, charge conductivity is larger and the variation of conductivity with temperature is more metallic. Organic superconductivity was first discovered in a Bechgaard salt, $(TMTSF)_2PF_6$ under pressure [Jérome *et al.* 1980]. One can put these two families of quasi-one-dimensional conductors in a generic phase diagram [Jérome 2012]. As seen in Fig. 8.2, the enhancement in the relative ratio of bandwidth to on-site repulsion (either by applying hydrostatic pressure or by the choice of the anion) leads to the emergence of metallicity and a superconducting ground state.

An extensive thermoelectric study by Mortensen and collaborators on Fabre salts clearly documented the evolution of thermoelectric response with emerging metallicity [Mortensen, Conwell, and Fabre 1983]. As seen I Fig. 8.3, a variety of behaviours can be seen in the data above 100 K. Among these compounds, $(TMTTF)_2Br$ stands out by clearly showing a thermoelectric response, which decreases with decreasing temperature. In the case of four anions ($ClO_4$, SCN, $NO_3$, and $BF_4$), the magnitude of the Seebeck coefficient is constant and close to 35 $\mu V/K$ in a wide temperature window. In two systems (($TMTTF)_2PF_6$ and $(TMTTF)_2AsF_6$), the high-temperature Seebeck coefficient displays a $1/T$ temperature dependence and tends to saturate to a value of about 30 $\mu V/K$. Upon cooling, at a given temperature (for example, $T_{MI} \sim 50\,K$ for

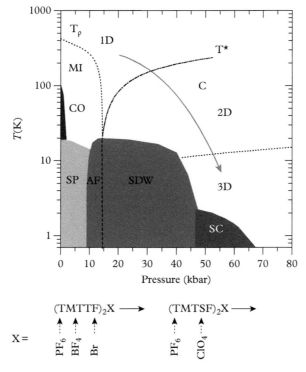

**Figure 8.2** *The generic phase diagram of Fabre and Bechgaard salts (after [Jérome 2012]. As one moves from left to right, electrons can hop more easily from one site to the other and metallicity strengthens. There is a variety of ground states including superconductivity, which emerges in the vicinity of a spin-density wave instability. The Fabre salts are at the right side of the Bechgaard salt, which implies that, at high temperature, they are more insulating than the Bechgaard salts. Moreover, (TMTTF)$_2$PF$_6$ is more insulating than(TMTTF)$_2$BF$_4$, which is in turn less metallic than (TMTTF)$_2$Br. Reprinted from [Jérome 2012] with kind permission from Springer Science and Business Media.*

(TMTTF)$_2$Br), the Seebeck coefficient deviates from linear or quasi-constant temperature dependence to attain a large magnitude of either sign. All these salts have an insulating ground state and the onset of the abrupt change in thermopower points to the onset of a metal–insulator transition.

Qualitatively, one can make sense of the high-temperature data. In (TMTTF)$_2$Br, the system closest to the emerging metallicity, electron hopping is strong enough to induce a metallic-like thermoelectricity. We saw previously that a quasi-linear Seebeck coefficient emerges if the chemical potential varies as $T^2$, which is the case of a

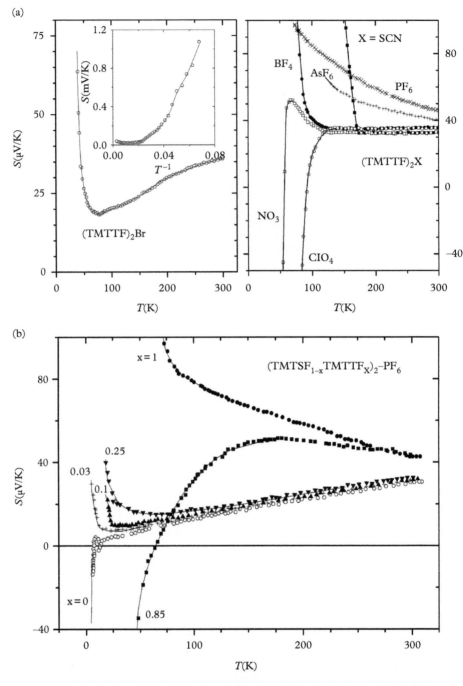

**Figure 8.3** *Seebeck coefficients in a number of Fabre and Bechgaard salts. (a) The Seebeck coefficient is metallic above 100 K in (TMTTF)₂Br and is temperature independent in a wide window in several other Fabre salts. (b) In Fabre-Bechgaard alloys, one can see the passage between metallic and semiconducting behaviours (adapted from [Mortensen, Conwell, and Fabre 1983] and [Mortensen and Engler 1984].*

Fermi-Dirac distribution. On the other hand, in $(TMTTF)_2PF_6$, the farthest from emerging metallicity, the Seebeck coefficient displays a semiconducting behaviour and varies as $1/T$, the behaviour expected when the chemical potential does not shift with temperature. It is easy to imagine that there can be an intermediate regime between the metallic and the semiconducting limit with a weak temperature dependence for the Seebeck coefficient. However, it is not clear why all the four salts should display a flat thermopower of $+35$ $\mu$V/K, very different from the flat $-60$ $\mu$V/K seen in several TCNF salts and interpreted as what should be expected in a strong-coupling Hubbard model. After all, neither of the two scenarios invoked above can explain a constant thermopower with a magnitude significantly different from $\frac{k_B}{e}ln(2)$. One may speculate that for some reason, the chemical potential shifts linearly with temperature giving rise to a constant Seebeck coefficient. The underlying reason for this linear shift remains obscure.

In this context, the results of another study on a Bechgaard-Fabre alloy [Mortensen and Engler 1984] are instructive. As seen in Fig. 8.3c, in the $(TMTSF_{1-x}TMTTF_x)_2$ $PF_6$ system, the passage between metallic and semiconducting responses can be documented. The Bechgaard salt, $(TMTTF)_2PF_6$ is a metal, at least above 11 K, below which it undergoes a spin-density-wave transition, driven by the nesting of the one-dimensional Fermi surface. As one replaces TMTSF with TMTTF, metallicity weakens and an insulating tendency emerges. At least up to $x < 0.25$ metallicity of the thermo-electric response persists. Above $x \geq 0.85$, on the other hand, the alloy becomes semiconducting. No flat temperature-independent thermopower was detected in any of the alloys, but may be at some intermediate $x$ between 0.25 and 0.85, one would have found one.

Thus, our current state of knowledge points to a somewhat disappointing conclusion on the issue of thermopower plateaus seen in a number of quasi-one-dimensional systems. They have been reported in at least two families and, in both cases, a number of distinct salts cluster around two well-defined number, which are fractions of $\frac{k_B}{e}$. We do not know, however, if this is due to a refined microscopic mechanism which leads to a $T$-linear shift in chemical potential or simply to a crossover between the two extreme limits of semiconducting and metallic behaviours.

Much attention has been paid to the fate of the electrons when a strong magnetic field is applied to Bechgaard salts and in particular to $(TMTSF)_2PF_6$ and $(TMTSF)_2ClO_4$. In this general context, there are a number of reports on the magnetothermoelectricity of these two salts.

If cooled slowly to allow its anions to order, $(TMTSF)_2ClO_4$ remains metallic at ambient pressure, and undergoes a superconducting transition. $(TMTSF)_2PF_6$, on the other hand, suffers a spin-density-wave instability at ambient pressure. Only the application of pressure suppresses this instability and allows the system to remain metallic before becoming a superconductor below a critical temperature of about 1 K. These quasi-one-dimensional conductors with a superconducting ground state both show a cascade of phase transitions induced by magnetic field. The ordered states have been identified as a set of field-induced-spin-density-waves (FISDW). The Hall coefficient becomes quantized in theses field-induced phases. This remarkable case of quantum

Hall effect in a bulk crystal has been ascribed to a nesting vector shifting continuously with magnetic field in order to keep the carrier concentration constant.

A study of Seebeck and Nernst coefficients in $(TMTSF)_2PF_6$ under pressure and subject to a magnetic field strong enough to induce the cascade of quantum Hall states [Kang *et al.* 1992] found that the thermopower becomes very large in this regime. A gap in the electronic spectrum is opened by these field-induced transitions. Therefore, a large Seebeck coefficient is not surprising. What is not understood, however, is the fact that, at the highest magnetic field (i.e. in the $n = 0$ state), where the system has an insulating charge conductivity, the Seebeck coefficient suddenly becomes vanishingly small. More generally, while the cascade of transitions does present visible signatures in the thermo-electric response, it has been hard to attain a quantitative picture of thermoelectricity in the FISDW regime of the Bechgaard salts.

Another axis of investigation has focused on effects arising from the fine structure of Fermi-surface sheets in the Bechgaard salts. The Fermi surface of these conduct-ors consists of two parallel sheets at $-k_F$ and $+k_F$. If these were truly one-dimensional conductors with no possibility for electrons to hope from one chain to the other, the two sheets would have been strictly featureless. But, this is not the case. Because of the existence of finite conductivity across the chains, the Fermi surface is warped. At high temperature, this small warping is not detectable because of the thermal fuzziness of the Fermi surface. This is no more the case as the system is cooled down and the sheets become three-dimensional objects in the reciprocal space. Lebed [1986] was the first to point out that, in such a context, there are specific angles defined by the crystal structure which reduce the dimensionality of the electron motion in the presence of a magnetic field. Numerous experimental studies have confirmed the existence of trans-port anomalies when the field is oriented along a magic angle. A comprehensive set of angle-dependent magnetoresistances can be found in a recent paper by Kang and co-workers [2007]. A recent account of the current theory can be found in a paper by Wu and Lebed [2010].

In both $(TMTSF)_2PF_6$ [Wu, Lee, and Chaikin 2003] and $(TMTSF)_2ClO_4$ [Choi *et al.* 2005], the thermoelectric response associated with a rotating magnetic field has been investigated. Remarkably, the most dramatic transport anomaly at a magic angle occurs in the transverse thermoelectric (i.e. Nernst) response (see Fig. 8.4). A satisfac-tory explanation of this experimental observation is still missing and we can only make a brief comment on the possible origin of the observed signal.

When the magnetic field is along a magic angle, because of the commensurability of two otherwise periods of oscillation, the dimensionality of electron motion in the semi-classical picture is reduced. This leads to a decrease in out-of-plane dissipation and a dip in $c$-axis resistivity. The Nernst anomaly, two extrema of opposite signs sandwiching this resistivity dip, is reminiscent of the thermoelectric response in a two-dimensional electron gas in the quantum Hall regime (see Chapter 10). Taking the extended Mott formula as a guide, the observed shape of the Nernst anomaly indicates that, at a magic angle, an infinitesimal shift of the chemical potential has no consequence on mobility, but very close to the magic angle, such a shift, would dramatically enhance a Hall signal.

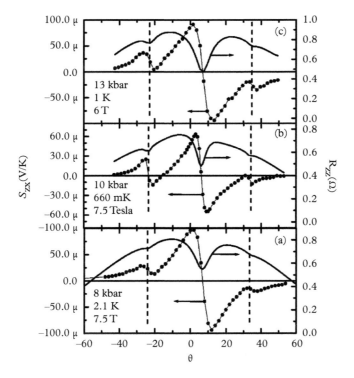

**Figure 8.4** *Out-of-plane magnetoresistance and Nernst coefficient for a rotating magnetic field in different temperatures and pressures in the metallic state of (TMTSF)$_2$PF$_6$. When the magnetic field is oriented along a magic angle, resistivity shows a dip and the Nernst coefficient presents a resonant anomaly consisting of two adjacent extrema of opposite signs sandwiching the minimum in resistivity. (After [Wu, Ong, and Chaikin 2005]).*

The opposite signs of the Nernst signal at the sides of the magic angle implies that only the component of the magnetic field, which is NOT along the magic angle, sets the response. A satisfactory microscopic scenario is yet to be found.

There are also several reports on thermoelectric studies on organic superconductors made from the BEDT-TTF molecules. These are two-dimensional conductors with a Fermi surface which has the shape of a distorted cylinder. Early measurements found that the in-plane Seebeck coefficient in $\kappa$-(BEDT-TTF)$_2$Cu(NSC)$_2$ has opposite signs when the thermal gradient is applied along two perpendicular directions [Urayama *et al.* 1988]. Deriving the band structure and the Fermi-surface topology from a tight-binding model, Mori and Inokuchi [1988] argued that this sign change can be traced to the structure of the Fermi surface, which is hole-like along one orientation and electron-like along the other. Their calculation reproduced the experimental data over a very wide

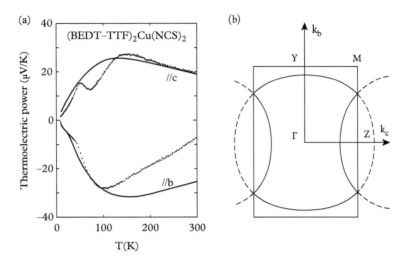

**Figure 8.5** *(a) Seebeck coefficient in κ-(BEDT-TTF)$_2$Cu(NSC)$_2$ for two different orients of thermal gradient has different signs. The expectations of a tight-binding model (solid lines) is in good agreement with the experimental data (solid symbols after [Urayama et al. 1988]. The sign of the Seebeck coefficient is different when the thermal gradient is applied along one or the other direction. (b) The calculated Fermi surface of the systems indicates that the curvature of the Fermi surface is hole-like along one direction and electron-like along another direction leading to different signs for the Seebeck coefficient. Reprinted with permission from [Mori and Inokuchi 1988]. Copyrighted by the Physical Society of Japan.*

temperature window, a rare achievement in the case of correlated metals (see Fig. 8.5). A similar sign anisotropy has been reported in the case of the sister compound κ-(BEDT-TTF)$_2$Cu[N(CN)$_2$]Br by Yu *et al.* [1991], who could also interpret their data by invoking the Fermi surface curvatures extracted from a tight-binding model. A similar sign anisotropy has been reported in the case of the Seebeck coefficient of another quasi-two-dimensional organic salt α-(BEDT-TTF)$_2$KHg(SCN)$_4$, indicating an anisotropic Fermi surface curvature [Choi, Brooks, and Qualls 2002].

There are a few studies devoted to the measurements of the Nernst coefficient in these two-dimensional superconducting salts. The Nernst signal associated with mobile vortices (see Chapter 9) was detected in both κ salts with Br and Cu(NSC) anions [Logvenov and co-workers 1997]. More recently, inspired by experiments on high-$T_c$ cuprates (see Section 8.3), the possible existence of a contribution by phase-fluctuating superconductivity to the Nernst signal above $T_c$ in these salts has been suggested [Nam 2007]. While the quality of data does not match what has been obtained either in cuprates or conventional superconductors (see Chapter 9), the study supports the existence of a contribution to the Nernst response by fluctuating superconductivity above $T_c$.

## 8.3  Cuprates

The discovery of superconductivity in a copper-oxide material [Bednorz and Müller 1986] led to a world-wide mobilization of thousands of condensed-matter physicists. Almost three decades and many bitter controversies afterwards, high-temperature superconductivity remains a puzzle with too many proposed solutions, none of them widely accepted.

The cuprate phase diagram is shown in Fig. 8.6. The parent compound of these materials (a solid with a chemical formula such as $La_2CuO_4$ or $YBa_2Cu_3O_6$) is an antiferromagnetic insulator, widely believed to be a Mott insulator. This is state of matter with a Coulomb interaction strong enough to invalidate the band picture. In the absence of a Coulomb interaction, the unit cell of $La_2CuO_4$ should host a single mobile electron, and therefore a large Fermi surface occupying half of the Brillouin zone is expected. Instead, it is an insulator and orders antiferromagnetically. The system can be p-doped by substituting lanthanum with strontium. By doping the system in this way, one creates a metal, and it is this particular metal which undergoes a superconducting transition. In almost all other cuprates, p-doping is achieved by introducing additional oxygen atoms, which each take two electrons out of the copper–oxygen plane, where all the action is happening (see Fig. 8.7).

Theoretical attempts to understand high-temperature superconductivity have been mostly framed in this context. The starting point is the doped Mott insulator [Lee,

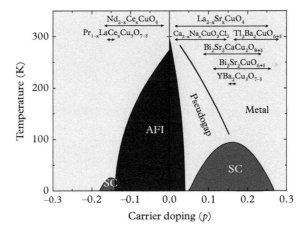

**Figure 8.6** *Generic phase diagram of electron-doped and hole-doped cuprates. The doping range open to exploration varies among different materials as shown in the inset. Reprinted from [Peets et al. 2007], © IOP Publishing Ltd and Deutsche Physikalische Gesellschaft. Published under a CC BY-NC-SA licence.*

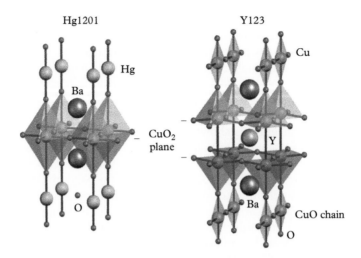

**Figure 8.7** *Left: Crystal structure of two cuprates. The conducting blocks are two dimensional and composed of adjacent basal planes of $CuO_6$ octahedra. The mercury-based $HgBa_2CuO_{4+\delta}$ (left) is single layer and doping is achieved by adding interstitial oxygen atoms. In $YBa_2Cu_3O_{7-\delta}$, each octahedron is cut in two, giving rise to a bilayer structure. Oxygen atoms added to $YBa_2Cu_3O_6$ occupy a site on Cu–O chains, introducing an additional conducting channel. Reprinted by permission from Macmillan Publishers Ltd: Nature [Barišić et al. 2013], copyright 2013.*

Nagaosa, and Wen 2006]. In 2008, superconductivity with a critical temperature almost as high as (but still somewhat lower than) in cuprates was discovered in a family of iron-based magnetic metals (for a review, see [Stewart 2011]). Following this discovery, the link between high-temperature superconductivity and Mott physics appears less direct than it did at the turn of the century. A recent radical theoretical approach puts in to question the very concept of a 'Mott insulator', making the undoped solid a magnetically ordered system which becomes metallic as soon as it is doped [Laughlin 2014]. Such an approach appears to be in conflict with the experimental fact that lightly doped cuprates above their Néel temperature are low-carrier conductors, as evidenced by the magnitude of their Hall coefficient [Ando *et al.* 2001] and also by the magnitude of their Seebeck coefficient, as we will see in the following pages.

Interestingly, the focus of the on-going dispute has not been the superconducting state by itself. It became soon clear that it hosts Cooper pairs, and experimental evidence accumulated that the superconducting gap vanishes along particular orientations. By the end of the 1990s a consensus had emerged that the superconducting order parameter has a *d*-wave symmetry [Tsuei and Kirtley 2000]. It is the metallic state, which has remained a subject of bitter controversy. Ironically, one persists in calling the latter 'normal', following the tradition established during the early decades of research on superconductivity.

There is no general agreement about how to describe this metal or how strange it is. If there is one consensus, it is about the conventionality of the overdoped regime, when superconductivity fades away with increasing carrier concentration. There is a widespread belief that, when hole doping exceeds the threshold to destroy superconductivity, the normal state becomes simple and understandable. Therefore, let us begin our survey of bulk transport in the cuprates by recalling the distinction between underdoped and overdoped limits.

As mobile holes are introduced in cuprates, antiferromagnetism fades at around a doping level of 0.05 holes per copper atom. Almost simultaneously with the destruction of the magnetic order, superconductivity emerges. The critical temperature steadily increases afterwards. This is the underdoped regime. Around $p \simeq 0.17$, the critical temperature attains its maximum value, which differs from one system to another (40 K in $La_{2-x}Sr_xCuO_4$ (LSCO) and 90 K in many others). A cuprate at this doping level is called optimally doped. Further doping leads to a decrease in the critical temperature and, by $p \simeq 0.33$, the metallic state persists down to zero temperature.

It is by now quite well established that the Fermi surface of overdoped cuprates is a large single-component, slightly warped cylinder. As expected from band calculations, this cylinder has a cross section which is $(1 + p)/2$ times the area of the Brillouin zone. In other words, at $p = 0.3$, it occupies about 65 per cent of the basal plane of the latter. This Fermi surface has been directly seen by angle-resolved photoemission (ARPES) [Peets *et al.* 2007] and also detected by quantum oscillations [Vignolle *et al.* 2008]. The available data on Tl-2201 indicates a perfect agreement between the two probes [Rourke *et al.* 2010]. Even before this convergence, transport properties of overdoped cuprates were found to be akin to what is expected in a Fermi liquid. The Wiedemann-Franz law has been verified [Proust *et al.* 2002] and the temperature dependence of resistivity becomes $T^2$ [Nakamae *et al.* 2003].

The underdoped side of the phase diagram has been the focus of most controversies. How does the large Fermi surface, the one expected in absence of Coulomb interactions, evolve as the carrier number is reduced? For a long time, one influential interpretation of the ARPES data suggested that the large Fermi surface was not shrinking in diameter but in circumference. As the system was underdoped, carriers were removed by wiping the Fermi surface along the so-called anti-nodal orientations, which ended up in a nodal metal with coherent quasi-particles only along the four nodal orientations. The main issue of controversy was about the nature of the pseudogap, a temperature scale below which a partial depletion in the density of states was observed by various experimental probes. Here, there were two major schools of thought opposing each other. On one side, the pseudogap temperature was considered a precursor of superconductivity. In an opposite picture, it was associated with another electronic order competing with superconductivity.

The observation of quantum oscillations in an underdoped cuprate [Doiron-Leyraud *et al.* 2007] had fatal consequences for this debate. This observation established that there is a small Fermi surface pocket (instead of a Fermi arc), at least, at low temperatures and in the presence of a magnetic field strong enough to destroy superconductivity. The recent observation of such oscillations in single-layer $HgBa_2CuO_{4+\delta}$ [Barišić *et al.* 2013] implies that this pocket is generic to underdoped cuprates and is not associated with copper–oxygen chains, a particularity of the $YBa_2Cu_3O_{7-\delta}$ system. The unexpected discovery of this small Fermi pocket changed the agenda. Now, the main issue appears to map the way it evolves to the large Fermi surface observed in the overdoped side. This is a formidable task. Superconductivity at optimal doping has proved to be too robust for detection of quantum oscillations in available magnetic fields. Even after this is figured out, it is far from clear that the answer would resolve the mystery of high-temperature superconductivity.

This section is devoted to a review of available data on thermoelectric response in cuprates. As detailed in other parts of this book, even in much simpler solids, which are fairly well understood, thermoelectricity presents a number of embarrassing and unresolved puzzles. It goes without saying that in the controversial field of high-$T_c$ cuprates, the unanswered questions outnumber by far the established facts. The good news is that the latter do not just boil down to the raw data. In the case of the small Fermi surface pocket in the zero-temperature limit, a rudimentary understanding of the Seebeck coefficient has emerged in the last couple of years.

The first measurements of the Seebeck coefficient of the hole-doped cuprates found a signal which was positive and almost temperature independent in an extended temperature range. Inspired by previous studies of the organic salts, this was interpreted in terms of the correlated hopping of the carriers in presence of strong Coulomb interactions. As we saw in Chapter 7, in such a Hubbard-based picture, the spin entropy of electrons is expected to give rise to a Seebeck coefficient of $\frac{k_B}{e} ln(2) \simeq 60 \ \mu V \ K^{-1}$, which would add up to a configuration entropy of $\frac{k_B}{e} \frac{1-x}{x}$, where $x$ is the number of carriers per site. However, early measurements did not find any variation of Seebeck coefficient with magnetic field, discarding the presence of the expected contribution of the spin entropy [Yu *et al.* 1988].

During the following years, extensive studies of the Seebeck coefficient were carried out in various families of cuprates and it became clear that the temperature dependence of the Seebeck coefficient has a structure, which evolves with doping (see Fig. 8.8 and 8.9). An early influential study [Obertelli, Cooper, and Tallon 1992] found that the magnitude of the Seebeck coefficient at room temperature is a measure of doping level $p$ across various families of cuprates. This 'OCT universal law' was particularly useful from a practical point of view. It is not straightforward to determine carrier density in cuprates other than LSCO. In the latter system, mobile holes are introduced by substituting lanthanum. In all others it is done by adding an amount of oxygen, which is difficult to quantify. Following the Obertelli, Cooper, and Tallon (OCT) law, the doping level could be determined by measuring the Seebeck coefficient at room temperature.

It soon became clear, however, that not all cuprates follow the OCT law [Ando *et al.* 2000; Dumont, Ayache, and Collin 2000]. More importantly, LSCO, the only cuprate in which chemical doping provides a strict measure of hole doping, does not follow the OCT law (See Fig. 8.9) [Cooper and Loram 1996]. On the other hand, a study of the Seebeck coefficient in $HgBa_2CuO_{4+\delta}$ [Yamamoto, Hu, and Tajima 2000] found that the parallel evolution of the room-temperature Seebeck coefficient and $T_c/T_c^{max}$ matches the expectations of the OCT law.

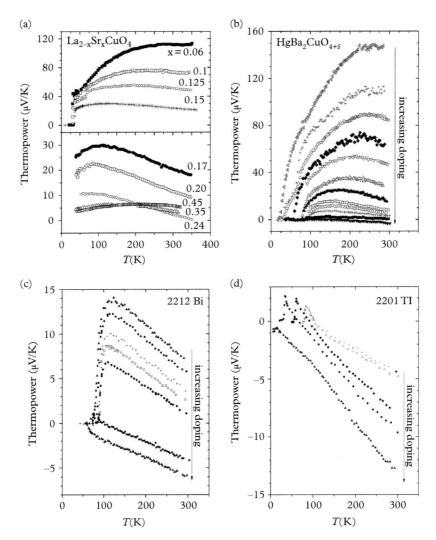

**Figure 8.8** *Evolution of the temperature dependence of thermopower with doping in various cuprates. (a) La$_{2-x}$Sr$_x$CuO$_4$ (adapted from [Cooper and Loram 1996]). (b) HgBa$_2$CuO$_{4+\delta}$ (adapted from [Yamamoto, Hu, and Tajima 2000]). Bottom right Bi$_2$Sr$_2$CaCu$_2$O$_{8+\delta}$ (Bi-2212) (adapted from [Obertelli, Cooper, and Tallon 1992]). (c) Tl$_2$Ba$_2$CuO$_{6+\delta}$ (Tl-2201) (adapted from [Obertelli, Cooper, and Tallon 1992]). Note that in the two latter systems, and in particular in Tl-2201, the doping variation is restricted to a finite window in the overdoped side.*

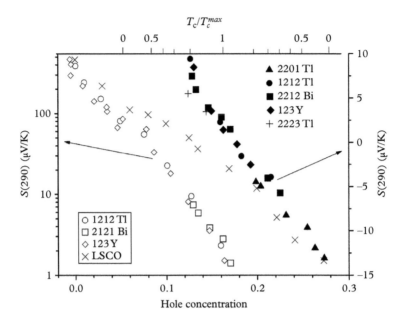

**Figure 8.9** *The Obertelli, Cooper, and Tallon (OTC) plot allows us to determine the doping level of various cuprates from their critical temperature and their room-temperature thermopower. The magnitude of the Seebeck coefficient at room temperature is plotted as a function of doping in several cuprates. Note the logarithmic vertical scale for the underdoped regime (y-axis at the right side) as opposed to the linear vertical scale for the overdoped regime on the left side. The data for $La_{2-x}Sr_xCuO_4$ (LSCO) (crosses), deviates from the OTC universal plot and remains positive in the overdoped regime (after [Cooper and Loram 1996]).*

The OCT plot is reproduced in Fig. 8.9. As seen in the figure, the magnitude of the room-temperature Seebeck coefficient steadily decreases with increasing doping. In the underdoped side, where the critical temperature is also decreasing with doping, this decrease is steep and seems to follow an exponential curve. At optimal doping, the room-temperature thermopower becomes as small as a few microvolts per kelvin. With further doping in the overdoped regime, the thermopower continues to decrease, much less drastically than in the underdoped side, following a linear dependence. As seen in the figure, the LSCO data do does not fall over those obtained for other cuprates. However, the trend is qualitatively similar. There is a steady decrease in the magnitude of the Seebeck coefficient, which becomes more moderate at higher doping level.

This phenomenological observation on the magnitude of the room-temperature See-beck coefficient can be roughly understood. Introducing a carrier to the system is expected to diminish the entropy per carrier, since lonely carriers have more degrees of freedom. The increase in carrier density enhances the Fermi energy $E_F$ and, since at a given temperature electrons can occupy available states within a thermally wide window of $k_B T/E_F$, the diffusive Seebeck response of hole-like carriers is expected to steadily decrease as new carriers are introduced by doping. It is difficult to go beyond this level of analysis with the data taken at the arbitrary temperature of 290 K and it is hardly surprising that the OCT law remains an approximate guide to determine the doping level.

What can be said about the temperature dependence of the Seebeck coefficient? One can recognize in Fig. 8.8 a pattern which smoothly evolves with doping. In underdoped samples, the positive thermopower increases almost linearly with temperature at first before saturating and becomes flat or slightly decreasing above a temperature of the order of $T^*$. In the case of $HgBa_2CuO_{4+\delta}$, it has been noticed that both the magnitude of $T^*$ and its doping dependence are close to the pseudogap line in the phase diagram [Yamamoto, Hu, and Tajima 2000]. Cooper and Loram [1996] have shown that, in the case of both LSCO and $YBa_2Cu_3O_{7-\delta}$, one can put all curves on top of each other using a scale in function using a $p$-dependent $S^*(p)$ and $T^*(p)$, both roughly tracking the pseudogap line.

Looking at these curves, one can hardly avoid thinking of the doped semiconductors reviewed in Chapter 7. In a degenerate Fermi liquid, the thermopower is proportional to $T/T_F$. As the temperature approaches the Fermi temperature, the increase of the Seebeck coefficient slows down and finally saturates. This scheme is visible in the lower panel of Fig. 7.2, a simple calculation of the expected response for p-doped and n-doped semiconductors as the Fermi temperature gradually shifts from 50 K to 900 K. Both the magnitude and the temperature dependence of the Seebeck coefficient in both LSCO and $HgBa_2CuO_{4+\delta}$ (Hg-2201) at, say, $p = 0.05$ is close but not identical to what one would expect for a Fermi liquid with a Fermi temperature of the order of 200 K, which is the expected Fermi energy of an electronic system with this carrier density and an effective mass close to the bare electron mass. It is reasonable to imagine that a good fit to the data can be obtained if one uses a characteristic energy scale evolving as a function of temperature. Thus, in the vicinity of the critical doping for the emergence of superconductivity (that is $p \simeq 0.05$), the thermoelectric response in cuprates is not very different from what one would expect in a Fermi liquid of such carrier concentration. It is useful to keep this in mind while debating the strangeness of the normal state in cuprates. This is another argument against Laughlin's picture of low-doped cuprates as high-density metals which would keep $1 + p$ carriers in the absence of magnetic ordering.

Before discussing the thermoelectric data quantitatively, let us underline the fact that, save for the case of $HgBa_2CuO_{4+\delta}$, the data in Fig. 8.8 were not obtained by studying single crystals and would smear out any anisotropy of the Seebeck coefficient. Measurements on single crystals have documented a modest yet significant anisotropy in thermoelectric response [Nakamura and Uchida 1993]. As seen in Fig. 8.10, the Seebeck coefficient in LSCO depends on the orientation of the applied thermal gradient. The observed anisotropy is not large, however. The Seebeck coefficient is positive for both in-plane and out-of-plane configurations, and the ratio of their magnitude remains well below 2, which is incomparably lower than the anisotropy of electric conductivity. The latter differs by more than two orders of magnitude along the two orientations of charge current. However, the mere existence of this anisotropy is telling

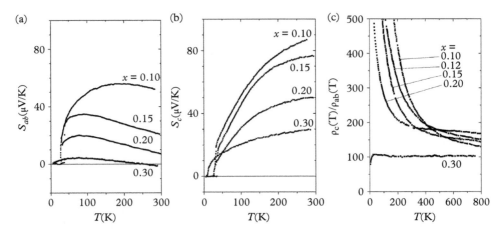

**Figure 8.10** *In-plane (a) and out-of-plane (b) thermopower in LSCO. The anisotropy of thermopower at a given doping level implies that the anisotropy of electric conductivity at that doping level is affected by an infinitesimal shift in the position of chemical potential. (c) The ratio of in-plane to out-of-plane resistivity in the same set of crystals confirms the evolution of anisotropy in electric resistivity with doping (adapted [from Nakamura and Uchida 1993]).*

us something important. The Seebeck coefficient is expected to be isotropic if it is governed by a single energy scale (often the Fermi energy, or an energy assimilated to it). This is true independently of the anisotropy of the charge conductivity. Indeed, since the anisotropy in the energy derivative of charge conductivity is compensated by the anisotropy of charge conductivity itself, the Mott formula would generate a Seebeck coefficient which remains isotropic no matter how large the anisotropy in charge conduction. In other words, if the anisotropy does not change with the shift in the chemical potential, the Seebeck coefficient is expected to be isotropic. This is indeed the case in a number of anisotropic Fermi liquids such as UPt$_3$ (see below). The anisotropy detected in thermopower suggests that, in cuprates in general and in LSCO in particular, the conductivity anisotropy is modified as the chemical potential shifts. Figure 8.10c shows that this is indeed the case [Nakamura and Uchida 1993] and this picture is qualitatively self-consistent. The larger out-of-plane Seebeck coefficient indicates a larger variation of out-of-plane conductivity with the shift in chemical potential.

Another point to keep in mind before any quantitative approach to the data is a possible role played by phonon drag. Such a contribution has been found to be sizeable in the overdoped regime of single-layer Bi$_2$Sr$_2$CuO$_{6+\delta}$ [Konstantinović *et al.* 2002], but still much lower than the diffusive component.

As mentioned above, the size and the shape of the Fermi surface on the overdoped side of the phase diagram appears to be established beyond reasonable doubt. In the case of Tl$_2$Ba$_2$CuO$_{6+\delta}$ (Tl-2201), thanks to extensive measurements of the dHvA effect [Rourke *et al.* 2010], we have a detailed knowledge of this Fermi surface. Close to $p = 0.25$, both the frequency ($F = \sim 18kT$) and the effective mass ($m_0 \sim 5m_e$)

have been determined. Therefore, the Fermi energy ($= \hbar^2 k_F^2$) is 4800 K. In the zero-temperature limit, the expected slope of the Seebeck coefficient should in this case be $S/T = 0.06$ $\mu$V K$^{-2}$. According to the early data reproduced in Fig. 8.8, extrapolating $S/T$ of Tl-2201 at this doping level to zero temperature would yield a value close to this. However, in the most overdoped samples of Tl-2201, the slope becomes negative, with an absolute value of $S/T$ wandering not very far from what is expected.

As seen in Fig. 8.8, in contrast to all other members of the cuprate family, LSCO keeps its positive low-temperature Seebeck coefficient up to very high values of doping. At $p = 0.3$, the slope of S/T in the zero-temperature limit remains positive and somewhat larger than expected ($\sim$ +0.1 $\mu$V K$^{-2}$). This value consistently emerges from the data reported by various authors [Cooper and Loram 1996; Elizarova and Gasumyants 2000; Nakamura and Uchida 1993]. Intriguingly, the Fermi surface of LSCO at this doping level resolved by ARPES becomes electron-like [Ino *et al.* 2002].

Tl-2201 and LSCO are the best-studied heavily overdoped cuprates and they differ also regarding their zero-temperature Hall coefficient. In overdoped Tl-2201 ($p \simeq 0.25$), the zero-temperature Hall coefficient is positive and its magnitude ($\sim$ +8 $\times$ 10$^{-10}$ m$^3$/C) corresponds to the expected carrier concentration of 1.3 holes per copper [Mackenzie *et al.* 1996]. On the other hand, in overdoped LSCO, the Hall coefficient is also positive (in spite of the electron-like Fermi surface) and much smaller than expected. For a detailed discussion on the magnitude and the temperature dependence of overdoped LSCO, see Narduzzo *et al.* [2008]. For a similar discussion on Tl-2201, see Kokalj, Hussey, and McKenzie [2012]. Thus, both the zero-temperature slope of the Seebeck coefficient and the magnitude of the Hall coefficient present non-trivial signs and magnitudes in the overdoped limit. The sign of the Hall coefficient is determined by the local curvature of the Fermi surface, integrated over the whole Fermi surface, with a local weight set by an eventually anisotropic mean-free-path [Ong 1991]. The sign of the Seebeck coefficient, on the other hand, is set by an energy derivative. It is positive if an infinitesimal upward shift in the chemical potential reduces conductivity.

Let us now turn our attention to underdoped cuprates. The way they were perceived was deeply modified by the discovery of quantum oscillations in YBa$_2$Cu$_3$O$_{7-\delta}$ [Doiron-Leyraud *et al.* 2007]. The mean-free-path of electrons in this system is much longer than in bismuth-based cuprates, which have been the subject of most ARPES studies, because of their cleavability. A long mean-free-path is the necessary condition for the observation of quantum oscillations. The reason this pocket was not detected by ARPES studies remains a subject of debate and speculation.

The results unambiguously established that, at least in YBa$_2$Cu$_3$O$_{7-\delta}$ and close to the doping level of $p = 0.12$, there is a small Fermi pocket. At this doping level, the Hall coefficient at low temperatures and high magnetic fields becomes negative, indicating that the carriers of the small Fermi pocket are electron-like [LeBoeuf *et al.* 2007]. Neither the location of this pocket in the Brilluoin zone nor the number of such pockets can be determined from quantum oscillations. However, the frequency of the oscillations (550 T) set the size of each pocket and the effective mass could be determined by the temperature dependence of the amplitude of the oscillations ($m^* = 1.76m_e$). The extracted Fermi energy from these two quantities was $T_F = 410$ K. The Fermi temperature sets

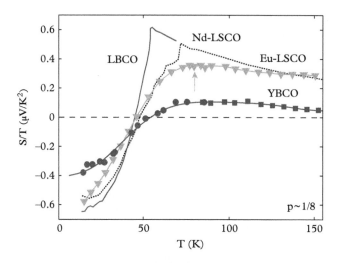

**Figure 8.11** *Thermopower divided by temperature, S/T, in several cuprates at p = 0.12, as a function of temperature. Reprinted from [Chang et al. 2010], copyright 2010 by the American Physical society.*

the expected magnitude of the Seebeck coefficient in the zero-temperature limit coming from pockets irrespective of their number and location [Chang 2010].

As seen in Fig. 8.11, measuring the Seebeck coefficient in $YBa_2Cu_3O_{7-\delta}$ at low temperature and in presence of a magnetic field large enough to destroy superconductivity leads to a quantitative determination of a finite negative $S/T$ of $-0.4$ $\mu V\,K^{-2}$. This is roughly half of the magnitude expected for the zero-temperature thermoelectric response of an electron pocket with a Fermi temperature of 410 K. It has been argued that the difference is compatible with the hypothetic presence of another undetected pocket with hole-like and low-mobility carriers. This remains a hypothesis, but a very plausible one. It is remarkable that the magnitude of $S/T$ in various cuprates at this doping level tends to be the same value as seen in Fig. 8.11 [Chang *et al.* 2010]. Quantum oscillations directly probing the structure of the Fermi surface were not detected in other cuprates. In their absence, this is the best available evidence for the presence of a small electronic pocket at this doping level in cuprates dirtier than $YBa_2Cu_3O_{7-\delta}$, as a consequence of the relative insensitivity to disorder of the Seebeck coefficient in the zero-temperature limit. The low-temperature Seebeck coefficient of cuprates have been found to drastically evolve with doping, and the largest negative $S/T$ emerges close to $p = 0.12$ in both $YBa_2Cu_3O_{7-\delta}$ and in europium-doped LSCO. Interestingly, the latter system is known to host at this doping level a particular type of electronic order which does not exist as a static long-range order in $YBa_2Cu_3O_{7-\delta}$. The similarity in the evolution of the zero-temperature thermoelectric response indicates that, in spite of this difference, the ultimate Fermi surface has the same structure in all cuprates.

Taking the universality of the low-temperature thermoelectric response in various cuprates as evidence for the ubiquity of the small pocket may appear speculative. However, it proved to be fruitful. First, it was found that in $HgBa_2CuO_{4+\delta}$ and near $p = 0.12$ at very low temperature, $S/T$ became negative and attained a magnitude comparable to other cuprates [Doiron-Leyraud *et al.* 2013]. Then, a few months later, quantum oscillations were finally observed in this system [Barišić *et al.* 2013], establishing that they were not exclusive to $YBa_2Cu_3O_{7-\delta}$ and its particular crystal structure. As this chapter is written in early 2014, Fermi surface reconstruction in various cuprates is a rapidly evolving subject of research and debate.

Much attention was given to the Nernst response of the cuprates. Following the observation of an anomalous Nernst signal in underdoped cuprates by Ong and co-workers in underdoped LSCO [Xu *et al.* 2000], the interpretation of the Nernst data occupied an important place in the debate on the origin of high-temperature superconductivity. The main experimental observation was the detection of a finite Nernst signal in the normal state of a number of underdoped cuprates. According to the influential interpretation put forward by the Princeton group, this signal was a manifestation of superconducting phase fluctuations in the pseudogap state of the cuprates [Wang, Li, and Ong 2006]. There were several reasons for the widespread popularity of this point of view.

First of all, Emery and Kivelson had argued in 1994 that a low superfluid density implies a low phase stiffness of the superconducting order parameter [Emery and Kivelson 1994]. Combine this observation with the assumption that the pairing interaction becomes stronger with the decrease in carrier density and you will find an appealing picture of a superconducting dome superposed on an additional pseudogap line. In this picture, superconducting transition on the overdoped side is governed by the pairing amplitude decreasing as a function of increasing doping level. On the underdoped side, in contrast, the transition takes place in two steps. First, below $T^*$, Cooper pairs are formed. Lacking phase coherence, they do not condense to the superconducting state before being cooled down to the critical temperature $T_c$, which marks the onset of the phase-coherent superconductivity. In this picture, the pseudogap is a precursor of superconductivity, and the temperature scale $T^*$ is associated with the amplitude of pairing interaction. Various experimental probes went under scrutiny in search of signatures of preformed but incoherent Cooper pairs between $T_c$ and $T^*$. After the experimental observation by the Princeton group in 2000, the Nernst data began to occupy a prominent location on this stage (see the review by Lee, Nagaosa, and Wen [2006]).

There was another reason for this widespread acceptance of the attribution of the observed Nernst signal to the persistence of large phase fluctuations of the superconducting order parameter in the pseudogap state. Back in the early oughties, the community knew little, and had forgotten much, about various sources of a Nernst signal. It was taken for granted that in metals, thanks to what was dubbed Sondheimer cancellation, the Nernst response is vanishingly small.[1] Moreover, the Nernst signal generated by Gaussian fluctuations of the superconducting order parameter has not been theoretically estimated

---

[1] 'The Nernst signal is generally much larger in ferromagnets and superconductors than in nonmagnetic normal metals' [Wang, Li, and Ong 2006]. The Nernst coefficient in semimetallic bismuth exceeds $1 \, mV \, K^{-1} T^{-1}$ [Mangez, Issi, and Heremans 1976]. This is four orders of magnitude larger than what was considered as anomalously large in underdoped LSCO.

and experimentally measured in other superconductors. Superconducting vortices were the only widely known source of a finite Nernst signal.

However, the landscape changed drastically during the following years. In various metals, a finite Nernst signal was measured. First, it was attributed to exotic physics in the context of strong correlation. Soon, however, it became clear that the Nernst signal is roughly proportional to the ratio of mobility to Fermi energy [Behnia 2009]. All strongly correlated metals, including those reported to host a 'giant' Nernst signal, were easily dwarfed by elemental bismuth. The emerging picture was that a small Fermi surface with high-mobility carriers is a natural source of a Nernst signal. What was found in $YBa_2Cu_3O_{7-\delta}$ in 2007 was exactly such a Fermi-surface pocket. The magnitude of the experimentally resolved Nernst signal in $YBa_2Cu_3O_{7-\delta}$ near $p = 0.12$ is very close to what is theoretically expected. Since the sign of the Nernst signal in this case is opposite to what is expected for vortices, the debate was settled, at least for this cuprate.

The Princeton discovery led to a parallel chain of events beginning with the elaboration of a theory for the Nernst signal generated by Gaussian fluctuations of the superconducting order parameter [Ussishkin, Sondhi, and Huse 2002]. This theory was experimentally tested a few years afterwards in a conventional dirty superconductor [Pourret *et al.* 2006]. A very good agreement between the experimental data on $Nb_{0.15}Si_{0.85}$ and the Ussishkin, Sondhi, and Huse (USH) theory was found. Thanks to a combination of large experimental resolution, low critical temperature (0.16 K) and low mobility of normal electrons, the Nernst signal generated by short-lived Cooper pairs could be followed up to 6 K (i.e. thirty times $T_c$). This demonstrated that the Nernst effect can indeed be a very sensitive probe of superconducting fluctuations. But, it also demonstrated that there is no need to have a pseudogap, strong correlations, or a doped Mott insulator to see fluctuating Nernst response in the normal state of a superconductor. The experiment on $Nb_{0.15}Si_{0.85}$ will be discussed in more detail in Chapter 9. A recent study of Nernst effect in cuprates [Chang 2012] led to the conclusion that, when the quasi-particle contribution is deduced, the remaining residual Nernst signal has an amplitude and temperature dependence close to the theoretical expectations of the theory for Gaussian fluctuations. Thus, in the beginning of the second decade of this century, evidence for phase fluctuating superconductivity in the pseudogap state of cuprates, compelling to many just a decade ago, has almost evaporated. One of the reasons for this paradigm shift is the way we look at the transverse thermoelectric response of both Fermi liquids and conventional superconductors.

In the past couple of years, Nernst measurements provided new input into this lively debate by resolving an unexpected in-plane anisotropy in the thermoelectric response. This anisotropy, which was found to emerge below $T^*$ (the temperature scale for the onset of the pseudogap state in the underdoped regime) was observed in orthorhombic yttrium barium copper oxide [Daou *et al.* 2010]. The anisotropy was detected by measuring the Nernst coefficient with the thermal gradient applied along two perpendicular crystal axes in conducting planes and vanishes in the zero-temperature limit [Chang *et al.* 2011]. The latter observation points to a possible non-diffusive origin. In the beginning of 2014, the link between this anisotropy and the suspected nematicity of the normal state is a rapidly evolving subject field of research.

## 8.4 Other Oxides

Besides cuprates, several other families of oxides have been subject to thermoelectric studies in the past two decades. This research has been motivated either by fundamental interrogations, by potential applications, or by a combination of both.

One of the discoverers of the large thermoelectric response of $Na_xCo_2$ [Terasaki, Sasago, and Uchinokura 1997], in an interesting short paper on the high-temperature thermoelectricity of oxide materials [Terasaki 2011], put forward an appealing argument for the usefulness of oxides as thermoelectric materials in a temperature range close to 1000 K. According to Terasaki, the thermoelectric response of these materials is governed by the Heikes formula, while their conductivity is in the Ioffe-Regel limit. This combination leads to a sizeable power factor and at high temperatures (i.e. around 1000 K), with a thermal conductivity reduced to $1 \ Wm^{-1}K^{-1}$, which, as we saw in Chapter 7, is comparable to what one finds in PbTe or $Bi_2Te_3$, a figure of merit close to unity is to be expected.

Indeed, a figure of merit of the order of unity has been observed in several oxides near 1000 K. In $Na_xCoO_{2-\delta}$ single crystals, $ZT$ was reported to become as large as 1.2 at 800 K [Fujita, Mochida, and Nakamura 2001]. In another cobaltate $(Ca_2CoO_3)_{0.7}CoO_2$ (with the so-called misfit structure), a $ZT$ of 0.87 has been reported at 973 K [Shikano and Funahashi 2003]. A comparably large $ZT$ was also reported in single-crystalline whiskers of $Bi_2Sr_2Co_2O_y$ [Funahashi and Shikano 2002].

However, it is doubtful that the dominant thermoelectric contribution in any of these three cases is rigorously explained by the Heikes formula. According to the experimental data, in all three of them, the Seebeck coefficient varies linearly in the temperature range of interest. This is in sharp contrast to prediction from the Heikes formula, that is, a flat Seebeck coefficient representing the number of configurations available to a hopping electron. Whatever entropy electrons are carrying in these materials at 1000 K, it cannot be merely a temperature-independent configurational entropy. This brings us to the fundamental component of this research.

A number of thermoelectric studies of oxides explored electron correlation with this sensitive probe. Among these oxides, those belonging to the $ABO_3$ family of perovskites (see Fig. 8.12) occupy a prominent place. By putting different metallic elements at locations A and B, one finds a dazzling variety of ground states. $BaTiO_3$ is a notorious ferroelectric insulator. $SrTiO_3$, on the other hand, is an antiferrodistortive (or a ferroelastic) insulator, which becomes superconductor upon doping. $LaTiO_3$ is an antiferromagnetic Mott insulator. $LaMnO_3$ is also an antiferromagnetic Mott insulator, but upon doping becomes an itinerant ferromagnetic metal displaying colossal magnetoresistance and hosts a peculiar orbital order.

$Na_xCoO_{2-\delta}$, which is not a perovskite, has also been extensively explored by the correlated-electron community. This compound consists of triangular $CoO_2$ layers separated by planes containing sodium atoms [Terasaki, Sasago, and Uchinokura 1997]. As the sodium content is varied, the $CoO_2$ planes are doped and a rich phase diagram appears [Foo *et al.* 2004], with an insulating state at a specific doping level. The system even displays superconductivity in a limited range of sodium concentration, provided

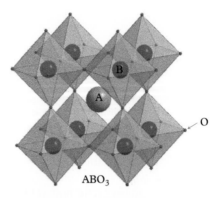

**Figure 8.12** *Perovskite crystal structure of the ABO$_3$ family consists of a network of vertex-sharing BO$_6$ octahedra. Each A atom is at the centre of a AO$_{12}$ cuboctahedron.*

that one dopes it with water [Takada *et al.* 2003]. In the next pages, we will briefly review thermoelectric experiments on this system as well as a number of perovskites. In spite of their potential applications, misfit cobaltites [Maignan *et al.* 2003] will remain out of the range of this review.

Figure 8.13 presents typical transport data for the Na$_x$Co$_2$O$_4$ system. The sodium concentration is close to the doping range at which the system displays a large Seebeck coefficient. For a metal of such carrier concentration, this is a large room-temperature Seebeck coefficient. Indeed, the Hall coefficient is measured to be several mm$^3$/C [Kawata *et al.* 1999] corresponding to a carrier density in the range of $10^{21}$ cm$^{-3}$, or a sizeable fraction (0.2–0.5) of one mobile carrier per formula unit. The combination of large Seebeck coefficient and moderate resistivity leads to a thermoelectric power factor of about 25 $\mu$W cm$^{-1}$K$^{-2}$ at room temperature [Fujita, Mochida, and Nakamura 2001]. This is comparable to the power factor of Bi$_2$Te$_3$, the best room-temperature thermoelectric material. However, the thermal conductivity is much larger, and only by going to 800 K can one find a competitive $ZT$ of 1.2, barely below the figure of merit in PbTe (reviewed in Chapter 7).

What is the origin of this remarkable thermoelectric response? According to calculations [Singh 2000], NaCo$_2$O$_4$ has a large Fermi surface. As a consequence of a very flat band dispersion, the density of states at the Fermi energy is large and, even in the absence of correlations, one expects an electronic specific heat as large as $\gamma = 21$ mJ K$^{-2}$. The measured value is two or three times larger ($\gamma \simeq 50$ mJ K$^{-2}$mol$^{-1}$); [Ando *et al.* 1999]), pointing to the presence of strong electron correlations. Thus, we are in the presence of moderately heavy electrons. With the heaviness set by a peculiar band dispersion combined with non-negligible electron correlations. In the isostructural compound LiV$_2$O$_4$, electron correlations become strong enough to make $\gamma$ eight times larger, transforming the system to the only known heavy-electron metal with no $f$ electron [Kondo *et al.* 1997].

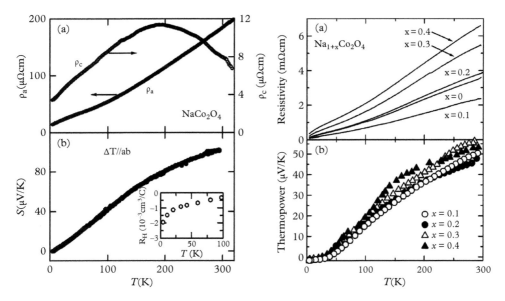

**Figure 8.13** *Left: Out-of-plane and in-plane resistivity and in-plane Seebeck coefficient of*
$NaCo_2O_4$ *(after [Terasaki, Sasago, and Uchinokura 1997]). Right: Variation with sodium content of*
*in-plane resistivity and thermopower (after [Kawata* et al. *1999]).*

Now, an electronic specific heat as large as 50 mJ $K^{-2}$ $mol^{-1}$ implies a large entropy
per volume. Assuming that there is one electron per formula unit, one expects a slope of
$S/T \simeq 0.5$ $\mu$V $K^{-2}$ for the Seebeck coefficient. This means a room-temperature Seebeck
coefficient as large as 150 $\mu$V $K^{-1}$, roughly twice what has been experimentally measured. Thus, the magnitude of the Seebeck coefficient by itself is not surprising. The
persistence of a large slope up to room temperature suggests that the principal source
of large thermoelectric response is the flat band dispersion, which pulls down the Fermi
energy to a mere 1000 K, leading to a room-temperature Seebeck coefficient of the order
of 100 $\mu$V/K [Singh 2000].

It has been argued that the likely source of enhanced thermopower in $Na_xCo_2O_4$ is
spin entropy [Wang *et al.* 2003; Lee *et al.* 2006]. Below 10 K, a drastic change in the
Seebeck coefficient induced by the application of a longitudinal magnetic field was observed [Wang *et al.* 2003]. The zero-field Seebeck coefficient (of the order of 1$\mu$ V/K,
at $T \simeq 5$ K), was found to decrease to virtually zero, at a magnetic field of 14 T. As we
saw previously, in a Hubbard picture of hopping electrons, the entropy electrons carry
is entirely due to their spin and the magnetic field, which lifts the spin degeneracy, is
expected to quench this entropy as well as the Seebeck coefficient caused by it. The
authors concluded that 'thermopower indeed derives entirely from the spin-1/2 excitations, and the complete suppression of [the Seebeck coefficient] reflects the removal of
the spin degeneracy by the applied field'. The problem with this interpretation is that
it leaves no place for charge entropy, the one measured by specific heat and giving a
fair account of the magnitude of the zero-field Seebeck coefficient. Given that all bulk

measurements document the existence of a strongly correlated Fermi liquid with well-defined quasi-particles and not hopping electrons, the relevant formula is Mott's and not Heikes'. Unfortunately, there is no report on an analysis of the field dependence of the Seebeck coefficient framed together with magneto resistance in this system.

Another interesting cobaltate is $LaCoO_3$, a distorted perovskite with a rhombohedral crystal structure and a non-magnetic insulating ground state. Substituting lanthanum with strontium or calcium introduces hole-like carriers and in $La_{1-x}Sr_xCoO_3$, a ferromagnetic ground state emerges when $x$ exceeds a critical level. This long-range ferromagnetism is concomitant with metal–insulator transition near the critical doping of $x = 0.17$, driven by percolation of ferromagnetic clusters already present for $x < 0.17$ [Wu and Leighton 2003].

The thermoelectric response of hole-doped $LaCoO_3$ has been the subject of several studies [Sehlin, Anderson, and Sparlin 1995; Berggold *et al.* 2005; Iwasaki *et al.* 2008]. As seen in Fig. 8.15, the thermoelectric response in the nominally undoped system is remarkably large at room temperature and is drastically reduced by warming. As seen in the right panel, there is also a dramatic enhancement of electric conductivity upon warming. The insulating ground state becomes conducting by passing through a thermal crossover, with no thermodynamic phase transition, in a manner similar to a narrow-gap semiconductor.

It has been suggested that warming introduces mobile carriers to this system, not by filling an unoccupied band but by charge disproportionation among trivalent $Ca^{3+}$ ions [Heikes, Miller, and Mazelsky 1964]. At high enough temperatures, a pair of $Ca^{3+}$ ions can become a divalent $Ca^{2+}$ and a tetravalent $Ca^{4+}$. This leads to the formation of polarons, carriers who have dug their own potential well by distorting the lattice through their electrostatic potential (see [Austin and Mott 1969]). These polarons are able to hop from one side to another, enhancing electric conductivity. As shown in Fig. 8.14, Sehlin, Anderson, and Sparlin [1995] demonstrated that, with such a picture and using the Heikes formula, one can quantitatively describe the temperature dependence of both the charge conductivity and the Seebeck coefficient. Assuming a constant energy gap of $\Delta G_D \simeq 0.38$ eV for the formation of divalent and tetravalent ions gives a rough description of the data. A much better fit can be found by assuming a concentration-dependent energy gap. With increasing polaron concentration $N$, one may expect the energy gap $\Delta G_D$ to decrease as a consequence of screening by the Coulomb potential as $\Delta G_D = \Delta G_0 - aN^{1/3}$. Indeed, as seen in Fig. 8.14 such a fit gives a good account of qualitative features in both $S(T)$ (a slight increase preceding the steep fall) and in $\sigma(T)$ (the belly preceding saturation at high temperatures).

Thus, $LaCoO_3$ warmed to above 600 K appears to be a rare case of undisputable relevance of the Heikes formula. The main mechanism of conduction appears to be hopping from one site to the other, leaving no other source of entropy for a charge carrier besides configuration. The Seebeck coefficient is almost temperature independent from 700 K to 1100 K, as expected from the Heikes formula. The evolution of the thermoelectric response with strontium doping (Fig. 8.15) is also compatible with this picture. As $x$ increases from 0.01 to 0.4, the high-temperature Seebeck coefficient of LSCO [Iwasaki

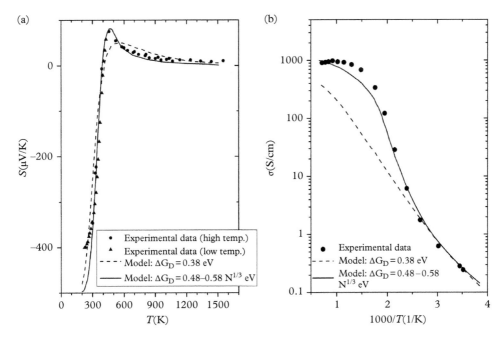

**Figure 8.14** *(a) Seebeck coefficient of nominally undoped LaCoO₃ as a function of temperature. (b) Electric conductivity of the same system. Solid (dashed) lines are theoretical expectation in a model with a constant (carrier-dependent) energy gap (after [Sehlin, Anderson, and Sparlin 1995]).*

*et al.* 2008] decreases gradually in magnitude, reflecting the variation in the number of available configurations, but remains flat as a function of temperature.

Below room temperature, the sign of the huge Seebeck coefficient of nominally undoped LaCoO₃ has been reported to be either positive or negative [Heikes, Miller, and Mazelsky 1964; Sehlin, Anderson, and Sparlin 1995; Berggold *et al.* 2005; Iwasaki *et al.* 2008]. This uncontrolled sample dependence points to infinitesimal concentrations of extrinsic impurities determining the sign. Controlled hole doping (by substituting lanthanum with strontium, for example) leads to a positive Seebeck coefficient which is reproducible and whose amplitude decreases with increasing doping, as seen in Fig. 8.15.

Another extensively studied member of the ABO₃ family is $La_{1-x}Sr_xMnO_3$ (For a review, see [Salamon and Jaime 2001]). A metallic ferromagnetic ground state emerges in this system upon doping. This order is believed to be driven by a double exchange interaction of electrons on the Mn site through Mn–O bonds, an idea put forward decades ago [Zener 1951]. In this picture, an oxygen atom can play the role of an intermediary in the process an electron travels from a $Mn^{3+}$ site to a $Mn^{4+}$ site through a $Mn^{3+}$–$O^{2-}$–$Mn^{4+}$ bond. Thus, electrons can be highly mobile without losing their spin orientation. This is the basic mechanism behind the itinerant ferromagnetism.

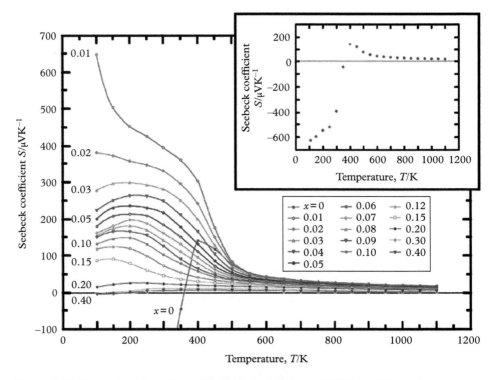

**Figure 8.15** *Temperature dependence of the Seebeck coefficient in LSCO for different doping levels x. Inset: data for the nominally undoped LaSrCoO₃. Reprinted from [Iwasaki et al. 2008], copyright 2008, with permission from Elsevier.*

The parent compound $LaMnO_3$ hosts an antiferro-orbital order, with the $d$ electrons of the Mn atom confined to a particular $e_g$ orbital, alternating across adjacent sites. This orbital order is accompanied with spin ordering, ferromagnetic in plane and antiferro-magnetic perpendicular to it [Imada, Fujimori, and Tokura 1998]. Upon doping with strontium, ferromagnetism emerges in $La_{1-x}Sr_xMnO_3$ for $x > 0.08$ and metallicity for $x > 0.16$ (see the left panel of Fig. 8.16). One motivation for studying this particular system has been it's 'colossal magnetoresistance'. Over a finite doping range, there is a phase transition from a high-temperature insulating paramagnet to a low-temperature metallic ferromagnet. In the vicinity of this phase transition, magnetoresistance is huge in magnitude. The transition from a high-temperature insulating state to low-temperature metallic state is accompanied by a drastic drop in resistance. This drop is drastically sensitive to the application of a small magnetic field; hence the large magnetoresistance.

The Seebeck coefficient of the system displays a strong variation with a magnetic field in the temperature range of colossal magnetoresistance. This can be seen in the bottom panel of Fig. 8.16, in particular in the sample where $x = 0.18$. The temperature dependence is complex and points to contributions with opposite sign evolving differently upon cooling.

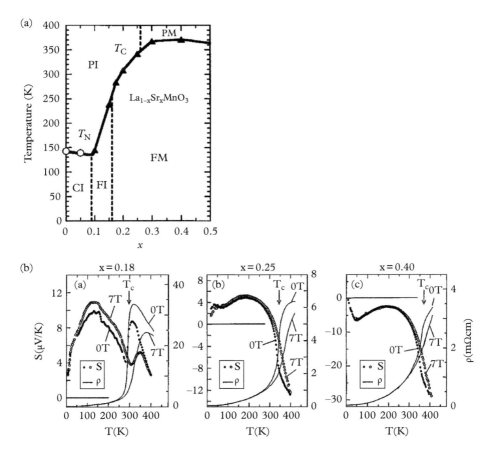

**Figure 8.16** *Top: Phase diagram of* $La_{1-x}Sr_xMnO_3$. *Reprinted from [Imada, Fujimori and Tokura 1998], copyright 1998 by the American Physical Society. Bottom: Resistivity and thermopower in three different crystals of* $La_{1-x}Sr_xMnO_3$ *with different doping levels. Reprinted from [Asamitsu, Moritomo and Tokura 1996], copyright 1996 by the American Physical Society.*

Well below the Curie temperature, the system has a metallic ground state. The residual resistivity is moderate, and inelastic resistivity is in $T^2$. The right panel of Fig. 8.17, shows the thermoelectric data on a calcium-doped sample with $x = 0.33$. One can see that the positive Seebeck coefficient of this metal with hole-like carriers is linear in temperature. The Fermi temperature extracted from this slope ($S/T \simeq 0.1\,\mu V K^{-2}$) is not very low ($T_F \simeq 2500$ K). It is in fair agreement with what is expected from the magnitude of electronics specific heat ($\gamma \simeq 4.7\,mJ\,mol^{-1}\,K^{-2}$) at this doping level. Both these probes points to a modest mass renormalization ($m^*/m_e \simeq 3.7$) [Salamon and Jaime 2001]. This is also compatible with the moderate amplitude of inelastic $T^2$ resistivity ($\simeq 5\,n\Omega\,cm\,K^{-2}$). Thus, far below the Curie temperature and deep inside the metallic state, the Seebeck coefficient appears understandable.

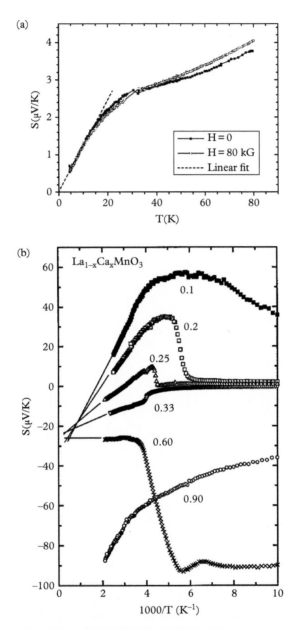

**Figure 8.17** *(a) Temperature dependence of the Seebeck coefficient in $La_{0.67}(Pb,Ca)_{0.33}MnO_3$. The low-temperature linear slope corresponds to the magnitude expected for a metal with the estimated Fermi energy of the system. Reprinted from [Jaime et al. 1998], copyright 1998 by the American Physical Society. (b) The Seebeck coefficient as a function of inverse of temperature in calcium-doped $SrMnO_3$ for different calcium contents. Reprinted from [Palstra et al. 1997], copyright 1997 by the American Physical Society.*

Many questions remain regarding the thermoelectricity of high-temperature insulating states [Palstra *et al.* 1997] (see Fig. 8.17b). The slope of the Seebeck coefficient plotted versus the inverse of temperature directly yields an energy gap which decreases with increasing doping. This feature is in qualitative agreement with what can be extracted from the activated behaviour of resistivity. However, the magnitude of the gap found from thermopower data is significantly lower, a feature attributed to ambipolar transport [Palstra *et al.* 1997]. This analysis failed to find a signature of the variation of the configurational entropy with doping, as expected from the Heikes formula in the thermoelectric data.

One of the most interesting members of the $ABO_3$ family of perovskites is $SrTiO_3$. It keeps its cubic structure down to 105 K and then undergoes a peculiar structural transition to a tetragonal structure. Each pair of adjacent $TiO_6$ octahedra slightly tilts clockwise and anti clockwise. This state is often called antiferrodistortive, to underline the opposite alignment of neighbouring tetrahedra, and sometimes ferroelastic, to underline the spontaneous loss of an elastic symmetry. At first sight, this is just a band insulator like its immediate $ATiO_3$ neighbours (A is an element residing in the same column with strontium). $BaTiO_3$, a celebrated ferroelectric, and $CaTiO_3$, not cubic even in temperatures exceeding 1000 K, are hard-boiled insulators with no detectable taste for metallicity.

Sandwiched between them, however, strontium titanate hosts a fascinating ground state at the crossroad of several sub-fields of condensed-matter physics. It has a very large dielectric constant, more than twenty thousand times larger than the dielectric constant of vacuum [Müller and Burkard 1979]. This 'quantum paraelectric' (dubbed so because ferroelectricity is avoided thanks to quantum fluctuations) can be easily transformed to a dilute metal. Infinitesimal n-doping, either by substituting titanium with niobium, or strontium with lanthanum, or by simply removing oxygen, leads to the emergence of a metal with extremely mobile carriers and subject to a superconducting instability, a property discovered as early as in the sixties [Schooley *et al.* 1964].

$LaTiO_3$ is also an insulator but is an antiferromagnetic Mott insulator. One can grow single crystals of $Sr_{1-x}La_xTiO_3$, with $x$ varying from 0 to 1. This is a unique opportunity to explore the properties of a system, which starts as a band insulator and ends as a Mott insulator, with a metallic state in between. In the nineties, the Tokura group performed such a study and found a metallic state with Fermi-liquid properties in a wide range of doping, extended up to $x \simeq 0.9$, above which the Mott insulator set in [Tokura *et al.* 1993]. In a subsequent study, almost a decade later [Ōkuda *et al.* 2001], the same group found that close to the band insulator (that is for $x$ close to 0), the system has a large Seebeck coefficient even at room temperature. Figure 8.18 presents their Seebeck and resistivity data. Inelastic resistivity follows a $T^2$ behaviour as expected in a Fermi liquid and the Seebeck coefficient is dominated by a large diffusive component with little phonon drag. As the band insulator is approached, the magnitude of both the $T^2$ resistivity and the $T$-linear Seebeck coefficient increases as a consequence of the decrease in the Fermi energy.

Early studies had already found that the magnitude of the room-temperature Seebeck coefficient in low-doped $SrTiO_3$ can approach millivolts per kelvin [Frederikse,

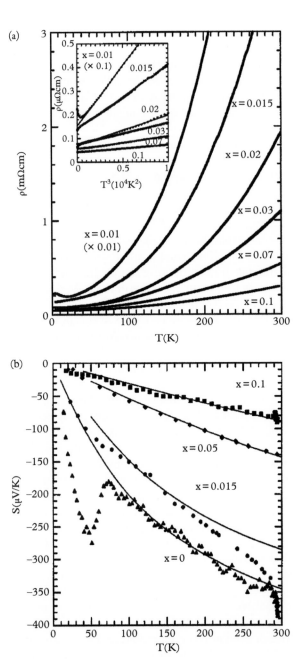

**Figure 8.18** *(a) Temperature dependence of resistivity in $Sr_{1-x}La_xTiO_3$. The system is metallic and resistivity follows a $T^2$ term as expected for a Fermi liquid. (b) The Seebeck coefficient of the same system is dominated by a large diffusive term, which is T-linear. The nominally undoped sample presents a phonon-drag peak. Reprinted from [Ōkuda et al. 2001], copyright 2001 by the American Physical Society.*

Thurber, and Hosler 1964]. Okuda *et al.* confirmed this finding and quantified the room-temperature power factor of $Sr_{1-x}La_xTiO_3$. In a narrow window of carrier concentration centred on $10^{21}$ cm$^{-3}$, it peaks to 35 $\mu$ W K$^{-2}$ cm$^{-1}$. This is comparable to the case of $Bi_2Te_3$. However, because of the large lattice thermal conductivity, exceeding 10 W K$^{-1}$ m$^{-1}$, the figure of merit remains below 0.1. Above room temperature, the situation improves a bit. At 1000 K, ZT was found to approach 0.4 in $Sr_{0.8}Nb_{0.2}TiO_3$ [Ohta *et al.* 2005], with a carrier density of $4 \times 10^{21}$ cm$^{-3}$. But, this is still much lower than PbTe, the reference material in this temperature range. The bottom line is that strontium titanate, lacking a heavy atom, has a disadvantageously large reservoir of lattice vibrations and can hardly compete with a compound with lead atoms such as PbTe.

A recent study of thermoelectric properties of lanthanum-doped thick films of $SrTiO_3$ with mobilities as high as 50000 cm$^2$ V$^{-1}$ s$^{-1}$ has documented the smooth evolution of thermoelectric response with doping [Cain, Kajdos, and Stemmer 2013]. As seen in Fig. 8.19, the Seebeck coefficient grows with decreasing concentration and a low-temperature phonon-drag peak gradually grows on top of the diffusive background. Note that as the doping decreases, the phonon-drag peak becomes more prominent and occurs at lower temperatures. We saw in Chapter 7 that the phonon-drag peak in bismuth occurs at 4 K, much lower than in noble metals. The phonon-drag peak is believed to occur at the temperature at which the typical wave-vector of thermally excited acoustic phonons matches the Fermi wave-vector of electrons. The evolution of the detected peak in $Sr_{1-x}La_xTiO_3$ is in qualitative agreement with this picture. In spite of the large power factor found at low temperatures, becomes of its large lattice thermal conductivity, n-doped $SrTiO_3$ has not been found a competitive thermoelectric material in cryogenic temperatures.

A study of thermoelectric response in n-doped $SrTiO_3$ extended to the sub-Kelvin temperature range made it possible to pin down the existence of a sharp Fermi surface with a simple topology at very low densities [Lin *et al.* 2013]. Giant quantum oscillations of the Nernst coefficient were resolved in $SrTiO_{3-\delta}$ samples. At a carrier concentration of $5.5 \times 10^{17}$ cm$^{-3}$, a single frequency of 18.2 T was resolved. This frequency made it possible to quantify the average radius of the Fermi surface, $k_F$. Together with the cyclotron mass, determined from the thermal evolution of the amplitude of quantum oscillations, this led to a quantification of the Fermi energy $E_F = \frac{\hbar^2 k_F^2}{2m^*} = 13\,K$. Figure 8.20 presents the data on quantum oscillations together with the zero-field, low-temperature Seebeck coefficient. As seen in the figure, $S/T$, which was around $40\,\mu\,VK^{-2}$ at 10 K (see Fig. 8.19b), extrapolates to $22\,\mu\,VK^{-2}$ at zero temperature. Injecting this value in the simple equation for the free electrons, $\frac{S}{T} = \frac{\pi^2}{3}\frac{k_B}{e}\frac{1}{E_F}$ yields 13 K for the Fermi energy, in excellent agreement with the analysis of quantum oscillations.

Such an experimental confirmation of the fundamental link between the magnitude of Seebeck coefficient and the Fermi energy is reminiscent of the case of metallic silicon (see Fig. 6.23 and the discussion). Thus, ironically, the most convincing cases of the validity of equation has not been performed in metals but in two semiconductors (silicon and strontium titanate) sufficiently doped to be on the metallic side of the metal–insulator transition.

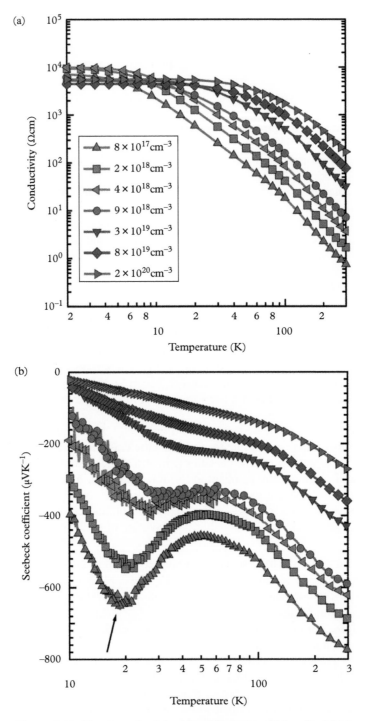

**Figure 8.19** *Electric conductivity (a) and Seebeck coefficient (b) of thick films of $Sr_{1-x}La_xTiO_3$. Reprinted with permission from [Cain, Kajdos and Stemmer 2013]. Copyright 2013, AIP Publishing LLC.*

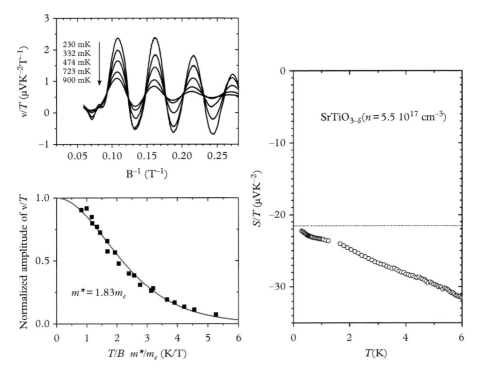

**Figure 8.20** *Quantum oscillations (right) and Seebeck coefficients (left) of SrTiO₃₋δ. The frequency of the quantum oscillation (top right) sets the average radius of the Fermi surface, while the temperature dependence of their amplitude sets the cyclotron mass. The extrapolation of S/T to zero temperature yields an unambiguous magnitude for the diffusive component of thermoelectric response. Both set of data lead to the same Fermi energy of 13 K (adapted from [Lin et al. 2013]).*

SrTiO₃ has a somewhat less well-known cousin. This is KTaO₃, which shares the same crystal structure and is another quantum paraelectric. A study of the thermoelectric properties of barium-doped KTaO₃ have found features rather similar to n-doped SrTiO₃ [Sakai *et al.* 2009]. For an electron concentration of $10^{20}$ cm$^{-3}$, the room-temperature figure of merit was found to be low ($ZT = 0.04$). Doped KTaO₃ has not been studied at very low densities yet.

Another explored member of the ABO₃ family, is the La₁₋ₓSrₓVO₃ alloy. In the case of this vanadate, LaVO₃ is the Mott insulator and SrVO₃ is a metal. The Seebeck coefficient has been found to evolve from a very large positive value near the insulator to a modest negative value near the metal [Uchida *et al.* 2011]. The metal–insulator transition occurs near a critical doping of $x = 0.18$. The evolution of the Seebeck coefficient with doping and temperature presents a complex structure with intriguing details. Nevertheless, Uchida and co-workers concluded that both the sign change and the magnitude of the measured thermopower are in agreement with the expectation of the Heikes formula.

## 8.5   Heavy-Electron Metals

As far as the author of these lines knows, the expression 'heavy-fermion quasi-particle' was first used by Steglich and collaborators in a paper reporting superconductivity in a ternary compound, $CeCu_2Si_2$, hosting such carriers [Steglich *et al.* 1979]. The discovery of heavy electrons, however is often dated to a few years earlier and the publication of a paper reporting a very large electronic specific heat in $CeAl_3$, which is neither superconducting nor magnetic [Andres, Graebner, and Ott 1975]. During the four following decades, many other heavy-fermion solids have been discovered and explored and there have been numerous studies of their thermoelectric properties. The first of such studies was reported as early as 1974, one year before the discovery of heavy electrons [van Aken *et al.* 1974].

The electronic specific heat of a Fermi liquid is linear in temperature. The prefactor of this $T$-linear specific heat, often dubbed $\gamma$ ($JK^{-2} mol^{-1}$), is proportional to the effective mass of electrons. Thus, the heaviness of electrons is quantified by the magnitude of $\gamma$. A material is qualified as a heavy-fermion system when $\gamma$ exceeds $100\ JK^{-2} mol^{-1}$, which is two orders of magnitude larger than the electronic specific heat of copper.

The physics behind the mass enhancement of electrons is the Kondo effect, which we already met when we discussed magnetic impurities in metals in Chapter 6. A localized magnetic moment put inside a Fermi sea of electrons is screened by the collective response of itinerant electrons. This is the original Kondo effect with subtle consequences, such as an upturn in resistivity and a large Seebeck coefficient. A heavy-electron metal is a Kondo lattice, a matrix of magnetic moments inside a Fermi sea. It contains an element such as cerium, ytterbium, or uranium with $f$-electrons, which provide the localized magnetic moments. The interaction between two adjacent magnetic moment favours a magnetic order which is usually antiferromagnetic. This antiferromagnetic coupling of moments mediated by itinerant electrons is the RKKY (after Ruderman-Kittel-Kasuya-Yoshida) effect. The Kondo effect governs the second type of interaction between each magnetic moment and the surrounding Fermi sea, which quenches the local moment. The heavy-fermion phenomenon emerges as a consequence of the rivalry between these two effects. If the RKKY coupling is strong enough, the system ends up by ordering with an antiferromagnetic ground state. If the Kondo effect happens to be stronger, then the system will remain a non-magnetic Fermi liquid of heavy quasi-particles. In some cases, this strongly correlated metal suffers a superconducting instability.

The microscopic mechanism behind this superconductivity appears to imply purely electronic degrees of freedom with no need for phonons as mediators. Direct evidence for this widely accepted hypothesis is still missing, but the fact that in most cases, heavy-fermion superconductivity is found in the vicinity of a magnetic quantum critical point (QCP) supports this picture. A QCP is a point which can be found on one axis of the phase diagram which separates two distinct states. The axis in question refers to an experimentally and tunable parameter such as pressure, magnetic field, or uniaxial stress, but not temperature. The phase transition between the two distinct ground states is a quantum phase transition. Quantum criticality refers to numerous signatures of fluctuating order parameter in the vicinity of such quantum phase transitions. These magnetic fluctuations are supposed to be the driving force behind the formation of Cooper pairs in heavy-fermion superconductors.

At room temperature, there are two kinds of electrons in a heavy-fermion metal. The $d$ electrons of the transition metal are those which carry charge. They are strongly scattered by localised magnetic moment associated with the $f$ electrons of the rare-earth metal. In this temperature range, the electric resistivity is flat or mildly increasing with decreasing temperature. As the system is cooled down, scattering is strengthened. Below a certain temperature, often called the coherence temperature, resistivity begins to drastically drop. The two kinds of electrons cease to be distinct and a new kinds of hybridized quasi-particle, heavy but mobile, is formed. Well below the coherence temperature, resistivity becomes quadratic in temperature with a large prefactor, as expected in the case of a Fermi liquid with very heavy quasi-particles.

The temperature dependence of the Seebeck coefficient in a heavy-fermion compound often presents a structure of intriguing complexity. Figure 8.21 presents a sample of early data. One can see that it is very different from one compound to the other. It is far from monotonous, with several peaks, more or less broad, and of either sign, superposing on each other.

Most theoretical attempts to describe the thermoelectricity of Kondo lattices have been based on the periodic Anderson model [Bickers *et al.* 1987; Schweitzer and Czycholl 1991; Costi and Hewson 1993; Mahan 1998; Zlatić *et al.* 2003]. In this model, there are two sets of electrons, itinerant and localized. The former are charge-carrying electrons of $d$ orbitals. The latter are localized $f$ electrons which are under the influence of an on-site Coulomb repulsion and which can hybridize with conduction electrons.

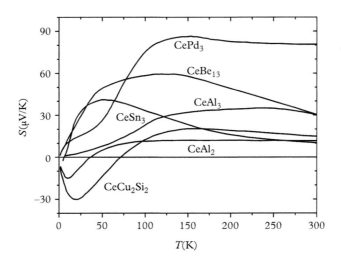

**Figure 8.21** *The temperature dependence of the Seebeck coefficient as a function of temperature in a variety of cesium-based heavy-fermion metals (after [Jaccard* et al. *1985]).*

This Hamiltonian has four distinct terms [Schweitzer and Czycholl 1991]:

$$H = \sum_{k,\sigma} \epsilon(k) c_{k,\sigma}^{\dagger} c_{k,\sigma}$$
$$+ E_f \sum_{R,\sigma} f_{R,\sigma}^{\dagger} f_{R,\sigma}$$
$$+ \frac{1}{2} U \sum_{R,\sigma} f_{R,\sigma}^{\dagger} f_{R,\sigma} f_{R,-\sigma}^{\dagger} f_{R,-\sigma}$$
$$+ V \sum_{R,\sigma} (c_{k,\sigma}^{\dagger} f_{R,\sigma} + f_{R,\sigma}^{\dagger} c_{k,\sigma})$$

(8.9)

In addition to the kinetic energy of the conduction electrons (the first term), there is a term which represents the typical energy of individual $f$ electrons, a term which quantifies on-site repulsion between two $f$ electrons of opposite spin occupying the same site, and, finally, a term which represents the possible conversion of a localized electron to a conduction electron and vice versa. The three relevant energy scales of the problem are $U$, $V$, and $E_f$. This is of course a crude approximation, since it does not take into account the orbital degree of freedom of the $f$ electrons. The crystal field environment should also be considered in any realistic treatment. However, the periodic Anderson model gives a picture of multiplicity of energy scales and their delicate balance (see [Mahan 1998] for a non-technical discussion). As the system is cooled down, the structure revealed in the temperature dependence of the Seebeck coefficient is a snapshot of this delicate balance. The relevant energy scales include the single-impurity Kondo temperature, the hybridization gap, the crystal field, and the original energy shift between $f$ and $d$ orbitals. As the temperature decreases, the number of relevant energy scales decreases and the behaviour becomes less complex and more generic.

The fact that the intricate structure found in the temperature dependence of the Seebeck coefficient is a consequence of such a delicate balance between multiple parameters can be seen in a striking manner by the application of hydrostatic pressure. Figure 8.22 presents one of such studies [Link, Jaccard, and Lejay 1996] on $CeCu_2Ge_2$. As seen in the figure, there is a smooth evolution of the thermoelectric response in the whole temperature range with pressure, leading to the disappearance of the negative peak around 20 K. A very similar evolution has been reported in the case of $CeCu_2Si_2$ [Jaccard et al. 1985]. Both these systems are heavy-fermion superconductors. However, while $CeCu_2Si_2$ superconducts at ambient pressure, one needs to apply a pressure exceeding 7 GP to make $CeCu_2Ge_2$ superconducting. The Seebeck coefficient of several other heavy-fermion systems has been studied under pressure and display a smooth evolution. To the best of our knowledge, in no case has the structure been quantitatively deciphered. We shall leave these Rosetta stones to future Champollions and limit ourselves to several general remarks.

The first question to address is the large magnitude of the Seebeck coefficient attained at room temperature. As one can see in Fig. 8.21, in some heavy-fermion compounds,

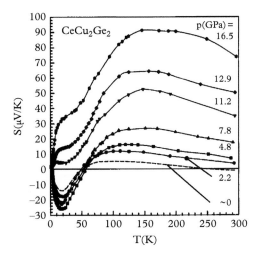

**Figure 8.22** *Evolution of the temperature dependence of the Seebeck coefficient with pressure in CeCu$_2$Ge$_2$. Reprinted from [Link, Jaccard and Lejay, 1996], Copyright 1996, with permission from Elsevier.*

thermopower can peak to a value as large as 100 $\mu\,\mathrm{V\,k^{-1}}$, two orders of magnitude larger than the typical values in simple metals. It has been noticed [Mahan 1998] that the power factor of many heavy-fermion compounds is larger than in Be$_2$Te$_3$ and they are disqualified as potential thermoelectric materials, only because of their high thermal conductivity.

   The question boils down to the following: where does the entropy come from? Each conducting electron in CePd$_3$ (Fig. 8.21) or in pressurized CeCu$_2$Ge$_2$ (Fig. 8.22) should have a lot of entropy to carry in order to see a Seebeck coefficient as large as a sizeable fraction of $\frac{\pi^2}{3}\frac{k_B}{e} = 286\,\mu\,\mathrm{V\,K^{-1}}$. In a dilute metal (such as a sufficiently doped semi-conductor), this happens because the low carrier density pulls down the Fermi energy and, at room temperature, there is a lot of phase space offered to mobile electrons by the thermal broadening of the Fermi-Dirac distribution. But, this is not what is happening in these systems, where there is roughly one electron per elementary crystal cell. At room temperature, these mobile electrons are not heavy either (they will become such only upon cooling), which would be another route towards pulling down the Fermi energy. In this respect, it is important to recall that the room-temperature electronic specific heat is not large in these materials, an observation which may appear puzzling at first sight. If there is a lot of entropy per *carrier* and if there is a high concentration of carriers, why doesn't one find a large specific heat, which is after all, a measure of specific heat *per volume*?

   The solution to this apparent puzzle is that the carriers acquire their entropy by getting scattered. The situation is akin to the case of single-impurity Kondo effect, which we saw Chapter 6. What enhances the Seebeck coefficient is the large dependence of the mean-free-path on energy. In the language of Eq. 5.25, the large Seebeck coefficient of CePd$_3$ at room temperature is mostly set by the second term of the right side (a large energy-dependent mean-free-path), while in Be$_2$Te$_3$, it is set by the first (a large thermal fuzziness of the Fermi surface).

The second point is the sign of the Seebeck coefficient. It is generally negative in ytterbium-based compounds and positive in cesium-based ones. This indicates that the energy derivative of the mean-free-path has opposite signs in these two sets of materials. In one case, the mean-free-path is strongly damped for carriers just beneath the Fermi surface and in the other, for those just above. This is in reasonable agreement with the Kondo resonance picture sketched in Fig. 6.12, assuming that cesium and ytterbium build Kondo resonances at opposite sides of the chemical potential.

With decreasing temperature, the Seebeck coefficient decreases before vanishing at zero temperature. As the thermal energy lowers below successive energy scales of the systems, the thermoelectric response becomes simpler. At the lowest temperature, the temperature dependence of the Seebeck coefficient presents a finite slope, which can be extracted by extrapolating $S/T$ to zero temperature. A decade ago, it was noticed that the slope of the Seebeck coefficient in various heavy-fermion metals correlates with the magnitude of the electronic specific heat [Behnia, Jaccard, and Flouquet 2004; Sakurai and Isikawa 2005] (see Fig. 8.23). Such a correlation would not be surprising if one pictures a heavy-electron system as a Fermi liquid with reduced Fermi temperature as a consequence of mass renormalization. Since specific heat and Seebeck coefficient both track the Fermi energy, they are expected to be enhanced in a similar way by strong correlation among electrons.

Fermi-liquid ratios relating two distinct signature of mass enhancement to each other are useful concepts. One of them is the Wilson ratio, which links the magnitude of magnetic susceptibility $\chi$ and the electronic specific heat $\gamma$ through a dimensional ratio:

$$R_W = \frac{4}{3} \left( \frac{\pi k_B}{\mu_B g} \right)^2 \frac{\chi}{\gamma} \tag{8.10}$$

In this expression, $\mu_B$ is the Bohr magneton and $g$ is the g-factor. These prefactors have been chosen such that $R_W$ becomes unity in a free-electron gas. In various strongly correlated electron systems, $R_W$ does not wander much above unity. This is not surprising, since an enhancement of density of states should lead to a parallel increase in both $\chi$ and $\gamma$.

Wilson ratio relates two thermodynamic properties of a Fermi liquid. The Kadowaki-Woods ratio, on the other hand, links a thermodynamic property, namely $\gamma$, to a transport property, that is, the inelastic quadratic resistivity $A$, in the expression $\rho = \rho_0 + AT^2$ for the resistivity of the Fermi liquids. Kadowaki and Woods [1986] found that, in various heavy-fermion metals,

$$R_{KW} = \frac{A}{\gamma^2} \sim 10^{-13} \ \Omega \ \text{m mol}^2 \text{K}^2 \text{J}^{-2} \tag{8.11}$$

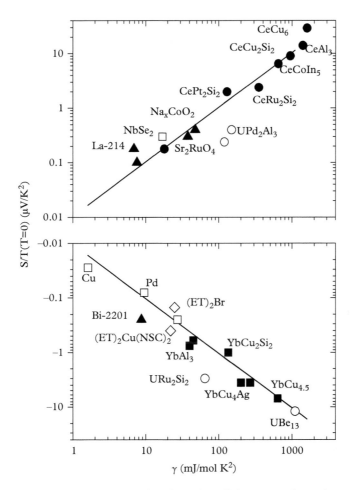

**Figure 8.23** *Initial slope of the Seebeck coefficient versus electronic specific heat in a number of correlated-electron systems. Reprinted from [Behnia, Jaccard and Flouquet 2004], © IOP Publishing. Reproduced by permission of IOP Publishing. All rights reserved.*

Here, the correlation arises because $A \propto \left(\frac{1}{E_F}\right)^2$ and $\gamma \propto \left(\frac{1}{E_F}\right)^2$. The Kadowaki-Woods correlation holds for numerous correlated-electron system and is a good guide to check if the measured magnitude of inelastic resistivity and electronic specific heat make sense. However, a close examination reveals that it is not universal. One can find a $R_{KW}$ ratio significantly different from the common value when the carrier concentration is very different from one electron per unit cell [Hussey 2005]. The magnitude of inelastic scattering, $A$, is model dependent and depends on the details of scattering. One should not forget that, the inelastic scattering of electrons from each other, first proposed by Baber [1937], is only expected when the Fermi surface is not a perfect sphere; thus, there is no such term for a free-electron gas.

The link between the Seebeck coefficient and the electronic specific heat is expressed through the dimensionless ratio [Behnia, Jaccard, and Flouquet 2004]

$$q = \frac{S}{T} \frac{N_{Av} e}{\gamma} \tag{8.12}$$

Here, $N_{Av}$ is the Avogadro number. This ratio is $-1(+1)$ for a gas of electrons (holes). As seen in Fig. 8.24., in many heavy-fermion metals, it is close to unity. Since the publication of the first observation of this correlation, quantitatively expressed as $q \simeq 1$, numerous experimental reports on various correlated systems found a $q$ close to unity. This is only expected when the system has the ordinary carrier density of one carrier per formula unit. In dilute systems, $q$ is expected to be large and roughly track the inverse of carrier concentration. On the other hand, when there is a large $\gamma$, but no mobile carriers, then $q$ is small. Only itinerant quasi-particles contribute to the Seebeck coefficient.

Contrary to resistivity, specific heat, and Pauli paramagnetic susceptibility, all of which always present the same sign, the Seebeck coefficient can be positive or negative. In a compensated multi-band metal, one would expect to see a cancellation of the contributions by hole-like and electron-like components of the Fermi surface. Thus, one interesting implication of the observed correlation is that the dominant contribution is skewed towards a hole-like portion of the Fermi surface in the case of cesium-based heavy fermions and an electron-like one in the case of ytterbium-based ones.

Early theoretical treatments of the Seebeck coefficient in the periodic Anderson model concluded that the slope of the Seebeck coefficient at low temperatures could not be simply set by $\gamma$. Instead, it was expected to be extremely sensitive to scattering details

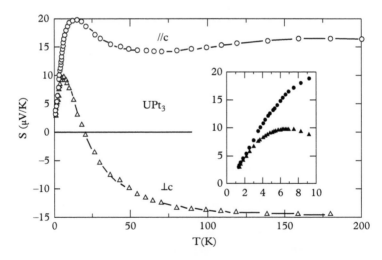

**Figure 8.24** *Temperature dependence of the Seebeck coefficient in single crystals of UPt₃. Inset is a zoom on the low-field temperature range. Reprinted from [Jaccard et al. 1992], © The Royal Swedish Academy of Sciences. Reproduced by permission of IOP Publishing. All rights reserved.*

and proportional to $\gamma \times \cot\theta$ [Costi and Hewson 1993; Zlatić *et al.* 2003]. Here, $\theta$ is the scattering phase shift, which can be anything between 0 and $\pi/2$. In such a picture, given the wild variety of values a cotangent can take, no universal or semi-universal correlation between the two quantities were expected. Soon after the experimental report [Behnia, Jaccard, and Flouquet 2004], however, Miyake and Kohno [2005] showed that such a correlation is compatible with the periodic Anderson model. This conclusion was shared by a later and much more detailed study [Zlatić *et al.* 2007].

We conclude this chapter with a brief review of thermoelectric studies on five emblematic heavy-fermion compounds. These are: i) $UPt_3$, a heavy-fermion superconductor with a Fermi-liquid normal state and a multi-component superconducting phase diagram; ii) $URu_2Si_2$, which undergoes a mysterious phase transition before suffering a superconducting instability; iii) $CeCoIn_5$, also a superconductor, but with a normal state presenting strong signatures of proximity to a quantum critical point; iv) $YbRh_2Si_2$, which orders antiferromagnetically at very low temperature, much studied because of its absent superconductivity quantum critical point; and finally v) $CeRu_2Si_2$, a Kondo lattice, which does not order down to zero temperature but displays a metamagnetic transition upon the application of a magnetic field.

The temperature dependence of the Seebeck coefficient in $UPt_3$ is presented in Fig. 8.25 [Jaccard *et al.* 1992]. This system is much explored because of the multiplicity of its superconducting phases (for a review, see [Joynt and Taillefer 2002]). Its normal

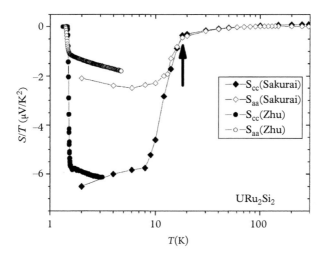

**Figure 8.25** *Temperature dependence of the Seebeck coefficient divided by temperature ($S/T$) in $URu_2Si_2$. High-temperature data are is taken from [Sakurai et al. 1996]. Low-temperature data from [Zhu et al. 2009]. The onset of hidden-order transition, marked by an arrow, is accompanied by a sharp and anisotropic jump in $S/T$, which persists down to the superconducting critical temperature.*

state, on the other hand, is an archetypal anisotropic Fermi liquid. Both in-plane and out-of-plane resistivity become $T^2$ below 1.5 K. As seen in the figure, the Seebeck coefficient has different signs for in-plane and out-of-plane temperature gradients at room temperature. The temperature evolution is also different for the two orientations. However, below 3 K, a Seebeck coefficient which is linear in temperature and isotropic in magnitude emerges. The slope of the Seebeck coefficient in this zero-temperature limit yields $S/T = 2.5 \, \mu\text{V}\,\text{K}^{-2}$ for both directions. Given that $\gamma \cong 0.44\,\text{JK}^{-2}\,\text{mol}^{-1}$, this yields $q \sim 0.6$. The isotropy of the Seebeck coefficient is remarkable, since resistivity in $UPt_3$ is anisotropic by a factor of 2.6 from room temperature to the superconducting transition. This anisotropy reflects the anisotropy of the Fermi velocity, set by band structure and independent of renormalization. In the simple picture linking the Seebeck coefficient to a Fermi energy, the anisotropy of the Fermi velocity is to be exactly cancelled by the anisotropy of the effective mass, and it should not affect the amplitude of the Seebeck coefficient along various orientations. This picture, which does not hold in many known systems, seems to be the case of $UPt_3$ below 3 K, in spite of its complex multi-component Fermi surface.

Undoubtedly, $URu_2Si_2$ has been one of the most explored heavy-fermion compounds during the last decade. It is also a superconductor, with a critical temperature of 1.5 K. However, the superconducting instability occurs in a very different context. Before becoming a superconductor, the system goes through a phase transition at 17.5 K with drastic signatures in almost all transport and thermodynamic properties. This phase transition was discovered in the eighties by three different groups, but it took a decade for researchers to find that there is something puzzling about this phase transition: in spite of the huge jump in specific heat, clearly signalling a second-order phase transition, no clear order parameter could be identified. Almost three decades after its discovery, this hidden order has remained a subject of ongoing research and debate (for a review, see [Mydosh and Oppeneer 2011]).

The Seebeck coefficient of $URu_2Si_2$ single crystals was first measured by Sakurai *et al.* [1996]. The data are presented in Fig. 8.25 in a semi-logarithmic scale. As seen in the figure, the onset of the phase transition at 17.5 K is accompanied by a drastic decrease in the magnitude of $S/T$ along both orientations. This indicates that the entropy per carrier in the hidden order is much larger than in the high-temperature state. We know from the anomalies observed in other transport properties (resistivity, Hall effect, thermal conductivity, and Nernst effect) that the hidden-order state is a low-density metal, in which most of the Fermi surface has been gapped. A rough estimation, based on the Hall data, would yield 0.05 carriers per formula unit. This, in its hidden-order state, $URu_2Si_2$ has the particularity of being a heavy-electron semi metal. Its Fermi energy is low, because of two distinct factors conspiring together: a low concentration and a large effective mass. There is a subtle point, often overlooked, regarding the thermodynamics of the hidden-order phase transition. According to the specific-heat data, $\gamma$ is larger in the high-temperature state than in the low-temperature one. On the other hand, as we can see, the absolute value of $S/T$ is larger in the ordered state. Taken together, these two facts suggest that, while the ordered state has less entropy per volume, it has more entropy per carrier. These two features are perfectly compatible, given that the number of carriers is reduced by one order of magnitude in the hidden-order state.

The figure shows also the low-temperature data [Zhu *et al.* 2009], which provide a reasonable match with the high-temperature one [Sakurai *et al.* 1996]. Before vanishing at the superconducting transition, the magnitude of $S/T$ is instructive about the ordered state. Contrary to $UPt_3$, the magnitude of $q$ is much larger than unity. Moreover, it is anisotropic. Since $\gamma \simeq 65\,mJK^{-2}\,mol^{-1}$, $q$ becomes as large as 10 for a thermal gradient along the $c$-axis. The anisotropy, which persists down to $T_c$ and is not wiped out by applying a magnetic field [Zhu *et al.* 2009], indicates that the magnitude of the Seebeck coefficient is not set by a single scalar such as the Fermi energy, as expected in a simple Fermi liquid.

The Fermi surface of $URu_2Si_2$ in its hidden-order state has been extensively studied by measurements of quantum oscillations. There are multiple components to it and even the most recent studies on high-quality single crystals may not have detected all its components [Hassinger *et al.* 2010]. However, all observed pockets are small in size and occupy a small fraction of the Brilluoin zone, which confirms that the system is indeed a low-density metal. Besides the large $q$, another consequence of low concentration is the large Nernst signal, which emerges below the hidden-order temperature in $URu_2Si_2$ [Bel *et al.* 2004a]. The magnitude of this signal can be attributed to the large mobility of carriers combined with their small Fermi energy.

Another heavy-fermion superconductor extensively studied in the beginning of the twenty-first century is $CeCoIn_5$. It was discovered in 2000 as a member of the 115 family of superconductors (for a review, see [Sarrao and Thompson 2007]). The superconducting transition occurs at $T_c \simeq 2.3\,K$. At this temperature, $\gamma$ is about $350\,mJK^{-2}\,mol^{-1}$ but is still increasing with decreasing temperature. Resistivity, far from being $T^2$, is $T$-linear down to $T_c$. All this indicates that the system is far from being a Fermi liquid. The formation of heavy quasi-particles is still unfinished when the superconducting transition occurs. The system is believed to be very close to an antiferromagnetic quantum critical point. This is confirmed by the fact that $CeRhIn_5$, its sister compound, is antiferromagnetic at ambient pressure and becomes superconducting by the application of pressure.

While, at the onset of its superconductivity, transport properties of $CeCoIn_5$ are anomalous, a Fermi liquid is established by applying magnetic field and destroying superconductivity. In other words, there is a field-induced quantum critical point: a Fermi liquid with increasing characteristic temperature emerges just above the superconducting upper critical field. Following an earlier study [Bel *et al.* 2004a], an intensive study of $CeCoIn_5$ [Izawa *et al.* 2007] has documented the thermoelectric response in this context. The data are shown in Fig. 8.26. The drastic evolution of the thermoelectric response near the superconducting upper critical field is striking. At zero magnetic field, the magnitude of $S/T$ is anomalously low compared to $\gamma$. With the gradual destruction of superconductivity with increasing magnetic field, $S/T$ displays a tendency to grow with decreasing temperature, Below 0.2 K and above 6 T, the magnitude of $S/T$ stabilizes to what is expected for a Fermi liquid with its measured value of $\gamma$ [Bianchi *et al.* 2003]. Putting the Seebeck and specific heat data side by side, one finds $q$ to remain

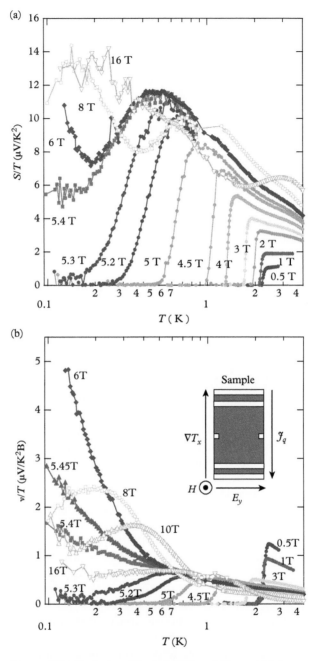

**Figure 8.26** *Evolution of Seebeck (left) and Nernst (right) coefficients in CeCoIn5 with the application of a magnetic field. Superconductivity is destroyed when the magnetic field exceeds 5.5 T; S/T and ν/T become large in this vicinity. For fields exceeding 6 T, a Fermi-liquid behaviour is recovered over a growing temperature window. Reprinted from [Izawa et al. 2007], copyright 2007 by the American Physical Society.*

close to unity near the quantum critical point. The Nernst coefficient shows a very large amplification near the quantum critical point and does not saturate down to the lowest temperatures investigated. Here, inelastic scattering, which remains prominent down to the lowest temperatures, appears to enhance the Nernst response.

$YbRh_2Si_2$ is another much-explored heavy-fermion system. It has a antiferromagnetic ground state with a very low Néel temperature of 70 mK. This antiferromagnetic order can be destroyed by the application of a very small magnetic field ($B_c = 60$ mT). In the vicinity of this critical field, the system has been found to show various signatures of quantum criticality (for a review, see [Gegenwart, Si, and Steglich 2008]). Hartmann *et al.* [2010] carried out a detailed study of the Seebeck coefficient of $YbRh_2Si_2$ and their data are reproduced in Fig. 8.27. At zero magnetic field, above $T_N$, $S/T$ was found to steadily increase with decreasing temperature. Its logarithmic divergence is reminiscent of the temperature dependence of $\gamma$ and also what was seen in $CeCoIn_5$. The onset of antiferromagnetic ordering leads to a sharp drop in $S/T$ at $T_N$.

As soon as a magnetic field larger than the small critical field is applied, a Fermi liquid is stabilized below a characteristic temperature, which increases with increasing magnetic field. In this temperature range, resistivity becomes $T^2$, and $\gamma$ becomes, constant. Hartmann *et al.* [2010] found that $S/T$ also saturates to a constant value below this characteristic temperature. The thermopower-to-specific heat ratio $q$ was found to remain close to unity in this Fermi liquid regime ($q = -1.2$ at $B = 0.1$ T and $q = -1.3$ at $B = 1$ T). Thus, in this system also, the recovery of the Fermi liquid is accompanied by the tendency of $|q|$ to become close to unity.

$CeRu_2Si_2$ goes through neither a magnetic nor a superconducting order, down to the lowest explored temperatures and has been dubbed the archetypal Kondo lattice (see [Flouquet 2005] for a review). When a magnetic field is applied along the *c*-axis of this tetragonal crystal, a metamagnetic transition occurs at $H^* \simeq 7.8$ T. This transition is associated with a jump in magnetization and is believed to be a true phase transition only at $T = 0$. At finite temperature, $H^*$ is only a cross over field scale, at which all transport and thermodynamic anomalies of the system show an anomaly. In particular, the magnitude of both electronic specific heat, $\gamma$, and inelastic resistivity, $A$, are enhanced at $H^*$. When the field exceeds $H^*$, the Kondo lattice is gradually destroyed and quasiparticles become light. Such a metamagnetic transition have been observed in numerous heavy-fermion materials, but the case of $CeRu_2Si_2$ is particularly well-documented. It is thought that, at zero temperature, there is a change in Fermi-surface topology at $H^*$, associated with a Lifshitz topological phase transition.

An early study of magnetothermopower in $CeRu_2Si_2$ [Amato *et al.* 1989] found that near $H^*$, the Seebeck coefficient presents a sharp anomaly and changes sign. A recent study [Pfau *et al.* 2012] found that the enhancements in $A$ and $\gamma$ are concomitant, so that the Kadowaki-Woods ratio remains constant (see Fig. 8.28). On the other hand, while the magnitude of $q$ remains close to unity both at zero-field and at high magnetic field, near the meta magnetic transition, $S/T$ and $\gamma$ display different behaviours as a function of magnetic field. Pfau *et al.* [2012] concluded that the result 'highlights the discrepancy between thermodynamic and transport properties across the metamagnetic

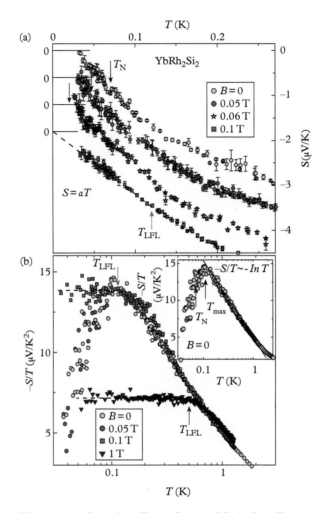

**Figure 8.27** *Seebeck coefficient(S) (a) and Seebeck coefficient divided by temperature (S/T) (b) in YbRh$_2$Si$_2$ near the field-induced quantum critical point. T$_{LFL}$ marks the temperature below which the system behaves as expected for a Landau Fermi liquid. Reprinted from [Hartmann et al. 2010], copyright 2010 by the American Physical Society.*

transition.' However, this does not explain why $\gamma$ correlates with a transport property (namely inelastic resistivity) and not with another (thermoelectricity).

One point missed in this analysis is the fact that electron-like and hole-like contributions compensate each other in the case of $S/T$ but add up for both $\gamma$ and $A$. We can assume two sources of electronic specific heat, one electron-like and the other hole-like, giving rise to the experimentally measured data of thermopower and specific heat.

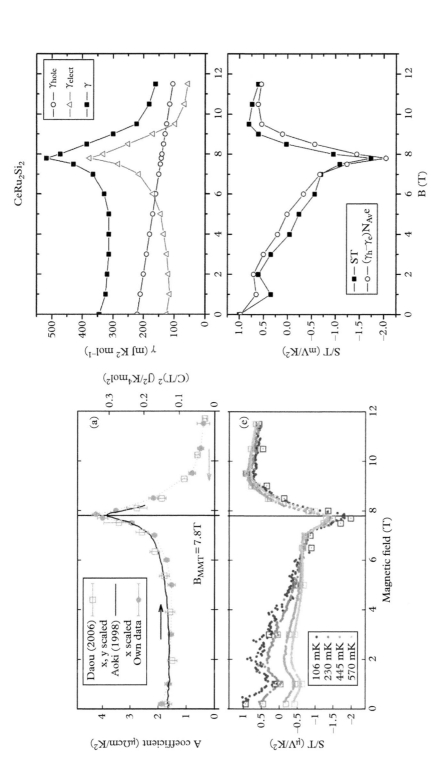

**Figure 8.28** *Left: Metamagnetic transition in CeRu$_2$Si$_2$ is accompanied by a peak in the magnitude of $T^2$ resistivity and electronic specific heat (top) as well as a sharp anomaly in S/T. Right: The experimentally measured values of $\gamma$ (top) and S/T (bottom) can be reproduced by taking a hypothetical $\gamma_e$ and $\gamma_h$, which add up to give $\gamma$ (top) and cancel out to give S/T (bottom). Intriguingly, only $\gamma_e$ shows a peak near H\*. Reprinted from [Pfau et al. 2012], copyright 2012 by the American Physical Society.*

The right panel of the figure indicates that one can easily find two hypothetic curves labelled $\gamma_e$ and $\gamma_h$, whose addition fits the measured $\gamma(B)$ in the whole field range and whose subtraction (assuming a conversion factor of $N_{Av}e$ reproduces measured $S/T(B)$). If this scenario happens to be true, the mass enhancement near $H^*$ should occur only in some (and not all) components of the multiple Fermi surfaces of the system. This would be an interesting feature of this metamagnetic transition, to be checked by further studies.

Doped with rhodium, $CeRu_2Si_2$ orders antiferromagnetically. In $Ce(Ru_{0.92}Rh_{0.08})_2$ $Si_2$, a magnetic field of $H_c \simeq 2.8\ T$ destroys antiferromagnetism, and metamagnetic transition occurs at a higher field of $H^* \simeq 5.8\ T$. A recent study of thermoelectricity in $Ce(Ru_{0.92}Rh_{0.08})_2Si_2$ has found an intriguing structure in both $S/T$ and $\gamma$ in the vicinity of both field scales, which is yet to be understood [Machida *et al.* 2013].

Three uranium-based heavy-fermion systems become ferromagnetic before going through a superconducting instability (for a review, see [Aoki and Flouquet 2012]). One of them is UCoGe, which is superconducting at ambient pressure. A study of thermoelectric response in UCoGe [Malone *et al.* 2012] quantified the magnitude of low-temperature $S/T$ in this system. As expected in this low-carrier system, the magnitude of the Seebeck effect was found to be large compared with that of specific heat ($q \simeq 5$), in rough agreement with our knowledge of the size of the Fermi surface in this system. This system also displays a metamagnetic transition upon the application of a magnetic field. A peak in $S/T$ in the vicinity of this transition has been observed and interpreted to the enhanced mass normalization near $H^*$.

In summary, while the order of magnitude of the Seebeck coefficient at very low temperatures in most heavy-fermion systems can be understood, a quantitative description of its variation with magnetic field and temperature remains a remote goal. Given the complex structure of the Fermi surface and our ignorance of the energy dependence of the scattering time, this is not very surprising.

# 9

# Superconductivity and Thermoelectric Phenomena

Many metals become superconductors upon cooling. Discovered in 1911, superconductivity has fascinated successive generations, before being explained by Bardeen, Cooper, and Schrieffer (BCS) in 1957. This is often considered as a major triumph of the quantum mechanics in the macroscopic world. According to the BCS theory, individual electrons cease to exist in the superconducting state, giving rise collectively to a macroscopic quantum wave-function with both an amplitude and a phase.

No electric field survives in a superconductor, at least in the absence of a magnetic field. If the superconductor is exposed to an external electric field, a dissipationless superfluid flow would annihilate the field. As a consequence, the Seebeck coefficient of a superconductor, measured as usual (i.e. by detecting the voltage created by a thermal gradient), becomes strictly equal to zero. Indeed, in a metal becoming superconductor, the Seebeck coefficient measured by experiment vanishes below the critical temperature as soon as the electric resistivity drops to zero.

However, this does not mean that thermoelectric phenomena are entirely absent in superconductors. At finite temperature, thermally excited quasi-particles are present and can carry entropy. Their flow is associated with an entropy flux and therefore there is a finite thermoelectric response (Fig. 9.1). Because of the presence of the superfluid background, however, this response cannot be detected with the standard set-up designed to measure the Seebeck coefficient of ordinary metals. This was pointed out as early as 1944 by Ginzburg (see [Ginzburg 1991] and references therein). There are a variety of thermoelectric phenomena observable in superconductors by unusual experimental set-ups, reviewed in detail by van Harlingen [1982].

There is a second type of thermoelectric phenomena associated with superconductors, which emerges only in the presence of a magnetic field. In a type II superconductor, an external magnetic field introduces superconducting vortices. These mesoscopic objects are filaments of magnetic flux associated with a whirling flow of charged electron pairs. Their motion generates a transverse electric field. An entire set of thermoelectric phenomena, mostly concerning the Nernst response, is associated with the dynamics of magnetic flux in the vortex state of type II superconductors.

*Fundamentals of Thermoelectricity*. First Edition. Kamran Behnia.
© Kamran Behnia 2015. Published in 2015 by Oxford University Press.

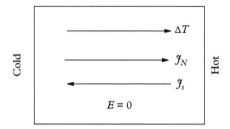

**Figure 9.1** *In a superconductor, a thermal gradient generates a flow of quasi-particles, $\mathfrak{J}_N$, compensated by a counterflow of superfluid, $\mathfrak{J}_s$. There is no finite electric field.*

Finally, a superconducting ground state has observable consequences for the thermo-electric properties of the metal becoming superconductor. Even above the critical temperature, when the condensate has been destroyed by thermal energy, traces of superconductivity can be detected in the thermoelectric response of the normal metal. This feature has been the subject of several experiments in the last few years.

In this chapter, we will review our theoretical understanding and available experimental evidence for three sets of thermoelectric phenomena. We will see that they are in very different stages of maturity. In particular, the gap between theory and experiment is quite different in the three research sub-fields.

## 9.1 Zero-Field Thermoelectricity in the Superconducting State

At absolute zero temperature, instead of being a set of individual entities as in the normal state, electrons of a superconductor form a many-body condensate. The 'Fermi sea' of the normal state is replaced by something even more fascinating, a system which responds to an external field like a single rigid body. In the temperature range between absolute zero and the critical temperature, it is often described with the so-called two-fluid model. Warming generates a normal fluid of quasi-particles that coexist with a superfluid background.

The superconducting transition opens a gap in the energy spectrum of independent electrons. The gap becomes larger upon cooling and then saturates to a finite value at zero temperature. A particularity of this gap is that it is pinned to the chemical potential and depopulates states just below or just above the Fermi level. In this respect, the superconducting gap is radically different from a band gap, whose locus does not correlate with the chemical potential. Thus, in a superconductor, electrons as individual particles can only exist when they are energetic enough to overcome this gap. The concentration of the normal fluid in the superconducting state is set by this.

We previously saw that the Seebeck response of a metal is generated by those electrons which are in the energetic vicinity of the chemical potential. Below the superconducting critical temperature, the gradual depopulation of states near the Fermi energy leads to a steady drop in the Seebeck coefficient. Nevertheless, the persistence of the one-particle spectrum beyond the gap implies a finite Seebeck response by thermally excited quasi-particles. As the temperature decreases, the thermal window, which quantifies the range of states which matter for the Seebeck response, sharpens. On the other hand, the gap, which quantifies the range of states excluded from occupation, enhances. As a consequence, the drop in Seebeck response becomes exponential at sufficiently low temperature (see Fig. 9.2).

As in the case of the normal state, the Boltzmann formalism can be employed to treat the transport properties of superconductors [Aronov 1981]. There is an important subtlety, however. Quasi-particles of the superconducting state differ from those in the normal state by their dispersion relation. Consider $\xi(k)$, the energy distance of an electron from the Fermi level in the normal state. It can be expressed as

$$\xi(k) = \epsilon(k) - \epsilon_F \approx \hbar v_F (k - k_F) \qquad (9.1)$$

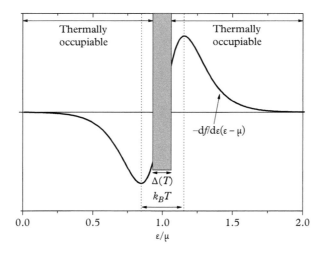

**Figure 9.2** *A gap opens in a superconductor and excludes states in the vicinity of the chemical potential from the one-particle spectrum. As the temperature decreases, the gap increases and the range of excluded states widens. On the other hand, the thermal window setting the range of those participating in the thermoelectric response sharpens. As a consequence, the thermoelectric response of quasi-particles in the superconducting state is drastically damped compared to their response in the metallic state at the same temperature.*

Such a linear approximation is valid if we restrict ourselves to the immediate vicinity of $\epsilon_F$. The slope of the linear dependence is set by the Fermi velocity. Note that the dispersion relation holds for both quasi-particles just above the Fermi surface (electrons) and those below (holes). In the superconducting state on the other hand, exciting the ground state generates the so-called Bogoliubov quasi-particles, which are combinations of hole-like and electron-like states. The dispersion relation for a Bogoliubov quasi-particle becomes [Ziman 1964]:

$$\epsilon^s(k) = \sqrt{(\xi(k)^2 + \Delta_0^2}$$ (9.2)

Such a quasi-particle is a linear combination of an electron and a hole. It significantly differs from normal quasi-particles in the immediate vicinity of the Fermi surface, as is illustrated in Fig. 9.3. In the normal state, thermal excitation continuously generates quasi-particles with a small energy difference with the Fermi level. This is no more the case in the superconducting state. The Fermi sea is frozen within a thickness of $\Delta_0$, the thickness of superconducting gap.

A route similar to the one that led to Eq. 2.39 can be taken in order to calculate the thermoelectric coefficient of the superconducting state $\alpha_s$. Compared to its value in the

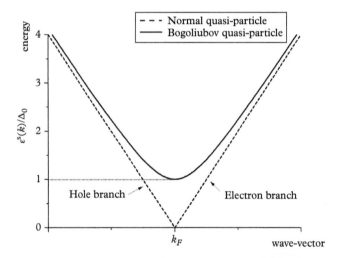

**Figure 9.3** *Quasi-particle dispersion is different in normal and superconducting states. In the normal state, in the immediate vicinity of the Fermi surface, there are excitable states whose energy and wave-vector is infinitesimally different from those located at the Fermi surface. In the superconducting state, however, quasi-particles closest to the Fermi surface are linear combination of electrons and holes. They tend to become like ordinary normal quasi-particles at high enough energies, when $\epsilon \gg \Delta$.*

normal state, $\alpha_n$, $\alpha_s$ is reduced by a temperature-dependent function:

$$\frac{\alpha_s}{\alpha_n} = G(\Delta/k_B T) \tag{9.3}$$

Galperin and co-workers [Galperin *et al.* 1974] calculated the expression for the function $G\left(\frac{\Delta}{k_B/T}\right)$. This function is similar, but not identical, to the normalized thermal conductivity of a superconductor, originally calculated by Bardeen and co-workers [Bardeen, Rickayzen, and Tewordt 1959]. The difference between the two expressions comes from the kernel in the integrals in the equations discussed in Chapter 2 (2.39 and 2.40). It is a linear function of energy for the thermoelectric response and it is quadratic in energy in the case of thermal response. In both cases, when the superconductor is cooled to temperatures well below its critical temperature, the expected signal becomes exponentially small. The function $G\left(\frac{\Delta}{k_B/T}\right)$ is defined by an integral [Galperin *et al.* 1974]:

$$G(x) = \frac{3}{2\pi^2} \int_x^\infty \frac{y^2 dy}{\cosh^2(y/2)} \tag{9.4}$$

This function has simple expressions in two opposite extremes. If $x << 1$, it becomes [Galperin *et al.* 1974]

$$G(\Delta/k_B T) \approx 1 - 2\left(\frac{\Delta}{\pi k_B T}\right)^2 \tag{9.5}$$

This corresponds to a gap much smaller than the thermal energy, a situation which occurs in the vicinity of $T_c$. As expected, at the critical temperature (when $\Delta$ vanishes), the thermoelectric coefficient recovers its normal-state value.

The other extreme, $x >> 1$, occurs at temperatures much lower than the critical temperature [Galperin *et al.* 1974]:

$$G(\Delta/k_B T) \approx 24\left(\frac{\Delta}{\pi k_B T}\right)^2 exp(-\Delta/k_B T) \tag{9.6}$$

This implies that the thermoelectric response of a fully gapped superconductor, like its electronic thermal conductivity and heat capacity, should exponentially vanish at very low temperatures. Basically, this reflects the rapid exponential decrease in the population of thermally excitable quasi-particles.

Thus, we have specific and quantitative theoretical expectations for the magnitude and temperature dependence of quasi-particle thermoelectricity inside the superconducting state. However, in contrast to other thermal properties, such as specific heat or thermal conductivity, the theory has not been tested by the experiment. As mentioned previously, the standard configuration, which measures a voltage drop in a sample submitted to a temperature gradient, cannot work. Cooper pairs will effortlessly compensate the quasi-particle signal and no finite voltage is expected along a superconducting sample subject to a thermal gradient.

Several alternative configurations have been proposed to distinguish between the flow of the superfluid and the normal current. One early idea, put forward by Ginzburg, is to study a superconductor with an anisotropic normal state. If the thermal gradient is not applied along a high-symmetry axis of such a system, because of the inherent anisotropy, the thermal gradient and the charge current would not be parallel to each other. This, combined to the fact that the magnetic field penetrates the sample within a finite penetration depth, leads to an observable magnetic field produced by the application of a thermal gradient along an arbitrary orientation. There is no record of an experimental attempt to fulfil this idea [van Harlingen 1982].

The most significant attempt to probe the thermoelectric response of a superconductor has been the bimetallic experiment carried out by van Harlingen and collaborators [van Harlingen, Heidel, and Garland 1980]. The results were found to be at odds with the theoretical expectations by a large factor. They have been the subject of a debate, which remains unsettled.

Consider a bimetallic superconducting ring consisting of a superconductor A and a superconductor B (Fig. 9.4). When the ring is cooled below the critical temperature of both superconductors, a macroscopic wave-function with a well-defined amplitude and phase, $\Psi = \Psi_0 exp(i\theta)$ will emerge. If a temperature gradient is applied along the two junctions of such a ring, the thermally induced quasi-particle flow in each ring is

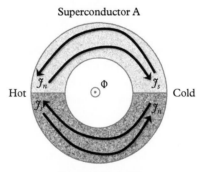

**Figure 9.4** *A bimetallic superconducting ring consists of two half-circular superconductors attached to each other. In the presence of a temperature difference between the two junctions, the quasi-particle flow $\mathcal{J}_n$ in each superconductor is compensated by a superfluid flow, $\mathcal{J}_s$. Phase coherence implies a finite magnetic flux set by circulating super current.*

compensated by a supercurrent. As a consequence, in each component of the ring, there will be a smooth spatial variation of the phase of the superconducting wave-function, $\theta$. This is allowed, provided the following constraint:

$$\oint \nabla\theta \cdot \vec{d\ell} = 2\pi n \qquad (9.7)$$

The constraint means that, save for an integer of $2\pi$, the macroscopic phase is well defined. This condition leads to the quantization of magnetic flux in a homogeneous superconducting ring with no temperature gradient:

$$\Phi = n\Phi_0 \qquad (9.8)$$

Here $\Phi$ is the magnetic flux through the ring and $\Phi_0 = \frac{h}{2e}$ is the quantum of magnetic flux. In the presence of a thermal gradient, however, the measured magnetic flux would show a deviation from its quantified value. This deviation would be proportional to the difference between the superfluid flow in component A and component B. Such a difference would be unavoidable, if the critical temperatures of the two superconductors happen to be very different.

In a series of experiments performed by van Harlingen and co-workers [van Harlingen, Heidel, and Garland 1980], seven different specimens of InPb rings were studied. The critical temperature of indium ($T_c = 3.4$ K) and lead ($T_c = 7.2$ K) substantially differ. The experimentalists had no difficulty to detecting signal. It was five orders of magnitude larger than what they expected. The magnitude of the signal was different among different samples, but appeared to scale with the normal-state thermoelectric response of indium in each case. Intriguingly, the signal was found to diverge in the vicinity of $T_c$ with a $(1 - T/T_c)^{-3/2}$ temperature dependence, instead of the expected linear behaviour (Fig. 9.5).

The surprisingly large discrepancy between the results of this experiment and the theoretical calculations led several authors to propose a number of solutions to the puzzle. Two authors [Marinescu and Overhauser 1997] argued against the application of the Boltzmann equation to superconductors. They developed a new 'electron conserving' transport theory which could generate the large thermoelectric signal observed by the experiment. However according to other authors [Galperin *et al.* 2002], this approach does not respect time-reversal symmetry. Moreover, putting in to question the applicability of the Boltzmann treatment of superconductors is problematic. As other authors have pointed out [Koláček and Lipavský 2005; Barybin 2008] the Bardeen, Rickayzen, and Tewordt theory of thermal conductivity in superconductors, which is based on such an approach, is compatible with numerous experimental results. Thus, rejecting Boltzmann's picture looks like throwing out the baby with the bath water.

An entirely different solution to this puzzle was proposed eight years later [Koláček and Lipavský 2005]. Koláček and Lipavský argued that since the superconducting penetration depth $\lambda$ is temperature dependent, the supercurrent density $\mathcal{J}$ can change along the temperature gradient as long as the product $\lambda\mathcal{J}$ remains constant. Moreover, in the

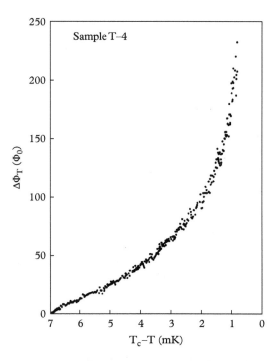

**Figure 9.5** *Magnetic flux generated in a bimetallic
InPb superconducting ring upon the application of a
thermal gradient. Reprinted from [van Harlingen,
Heidel and Garland 1980], copyright 1980 by the
American Physical Society.*

vicinity of the critical temperature, the superfluid density is much smaller than the total electron concentration. They argued that the magnetic flux generated by the temperature gradient due to the temperature dependence of $\lambda$ is by far larger than the one produced by the supercurrent compensating the thermally driven quasi-particles and indeed in good agreement with the experiment.

A more recent proposal has been a new Fermi liquid approach to the thermoelectricity of superconductors [Barybin 2008]. According to Barybin, the missing point in Galperin's theory is the failure to take into account the superfluid motion and to focus on the phase velocity of electrons instead of their group velocity. This proposal leads to a fundamental correction to the previous theory. Barybin finds a new 'semiconducting-like' term in the thermoelectric response, a term that had been absent in the previous treatments. This term, much larger than the original 'metallic-like' is argued to give an adequate account of the experimental result [Barybin 2008].

Thus, the treatment of the thermoelectric response in superconductors remains an open question. From the experimental point of view, the number of studies remains meagre and it is certainly desirable to extend these studies to superconductors other than indium and to use techniques other than the bimetallic set-up. From a theoretical point of view, it appears that a consensus on the adequate formalism to use has not been attained yet.

## 9.2   The Nernst Response of Mobile Vortices

There are two types of superconductors. This is because a superconductor has two relevant length scales: the coherence length and the penetration depth. The first one quantifies the average size of the Cooper pairs. The second one quantifies how deep a magnetic field can penetrate the superconductor. In type II superconductors, in contrast to the type I, the coherence length is shorter than the penetration depth.

An intriguing consequence of this inequality was theoretically discovered by Abrikosov in the fifties and was later confirmed by experiment. When the magnetic field exceeds a critical threshold, known as the first critical field, $H_{c1}$, it penetrates the superconductors as tiny filaments each carrying a quantum of magnetic flux. These mesoscopic objects known as Abrikosov vortices have an internal structure. They have a normal core, which extends over the coherence length, and a periphery of whirling Cooper pairs, extending over the penetration depth.

Superconductivity is effectively destroyed in the vortex core. The magnetic field also peaks at the centre of the vortex. As one moves from the centre to the periphery of a vortex, the superconducting order is rapidly restored and, with a milder pace the magnetic field vanishes. Of a superconductor exposed to a magnetic field larger than the lower critical field but lower than the upper critical field, $H_{c2}$, one says that it is in the mixed or vortex state. Since each vortex carries a finite amount of magnetic flux, the vortex population and the intervortex distance are uniquely set by the magnetic field. At sufficiently high magnetic field, the intervortex distance becomes comparable to the size of the vortex core. Above this field, the superconductor is totally occupied by the cores of the vortices and thus superconductivity is effectively destroyed. This is often called the orbital limit of superconductivity.

Vortices repulse each other and in the absence of disorder, they can form a triangular lattice. In the presence of disorder, the lattice is distorted. In the presence of a larger magnetic field, the vortex lattice or the vortex glass melts and vortices become mobile. Mobile vortices are subject to a thermal gradient drift, because in a superfluid environment they are entropy reservoirs. The normal core has more entropy than the superconducting background. Thermodynamics generates a thermal force pushing the vortex from hot to cold and this movement generates an electric field. A natural question arises: why don't Cooper pairs screen this electric field like the one caused by quasi-particles?

In order to see why, one has to consider another peculiar feature of a vortex. It is not only a carrier of both magnetic flux and entropy, but also a purely quantum-mechanical object. The superconducting wave-function has an amplitude and phase. As mentioned above, this wave-function totally vanishes in the centre of the vortex. Its amplitude becomes zero and its phase undefined. As one moves from the core to periphery, the wave-function acquires an amplitude and a phase. The amplitude depends on the radial distance and the phase on the azimuthal angle. Because there is a singularity in phase at the heart of the vortex, its displacement implies a change in the local value of the phase of the wave-function. This is what generates a transverse electric field. The magnitude of the electric field produced in this way is exactly the same that one would expect for a moving flux, but its origin is quantum mechanical. As a result, Cooper pairs accommodate with the electric field produced by vortex movement.

This electric field is a consequence of the reorganization of the condensate and not a perturbation external to it, like the one generated by normal quasi-particles.

Thus, in a superconductor with mobile vortices, one expects that a longitudinal thermal gradient generates a transverse electric field, in other words, a finite Nernst signal. The Bridgman relation between Nernst and Ettingshausen coefficients, which we saw early in this book, implies that, if there is a finite Nernst signal, there should also be a finite Ettingshausen effect. This is easily conceivable. If a current density of $\mathbf{J}$ flows in a superconductor, each vortex, carrying a magnetic flux of $\Phi_0$, would feel a Lorenz force equal to

$$F_L = \mathbf{J} \times \Phi_0 \qquad (9.9)$$

If the vortices are mobile, they move as a consequence of this Lorenz force and carry their entropy, generating a thermal gradient along their trajectory and perpendicular to the current density vector and, hence, an Ettingshausen signal.

Early experiments, performed on magnetothermoelectricity of type II superconductors in their mixed state, detected both Nernst [Lowell, Munñuz, and Sousa 1967; Vidal 1973] and Ettingshausen [Solomon and Otter (1967)] effects (Fig. 9.7). Such experiments are reviewed in an article [Huebener 1972] and a book [Huebener 1979] by Huebener.

A new era in the field of superconductivity was inaugurated by the discovery of high-temperature superconductors in 1986. In these superconductors, with a critical temperature exceeding the boiling temperature of nitrogen, vortices are mobile in an extended region of (field, temperature) plane. Following a pioneer experiment detecting the Ettingshausen signal [Palstra *et al.* 1990], numerous experiments on thermally induced movement of vortices in these family of superconductors followed, which are reviewed in another article by Huebener [1995].

The Nernst signal generated by superconducting vortices as they are displaced by an applied thermal gradient is described phenomenologically by considering the forces exerted on the vortices. The first one is the thermal force exerted by the thermal gradient:

$$\mathbf{F}_{th} = -\nabla T S_\phi \qquad (9.10)$$

Here, $S_\phi$ is the entropy transported by a single vortex. Moving vortices with speed $\mathbf{v}$ are also subject to a frictional force:

$$\mathbf{F}_f = \eta \mathbf{v} \qquad (9.11)$$

The damping viscosity is represented by $\eta$. Its presence ensures that vortices do not accelerate indefinitely. In a steady-state Nernst experiment, the frictional force balances the thermal force and thus, when the thermal gradient is applied along the x-axis, the vortices move with the following velocity:

$$\mathbf{v_x} = \frac{-\nabla_x T}{S_\phi}\eta \tag{9.12}$$

The movement of flux lines with such a velocity would generate an electric field perpendicular to their velocity (along the x-axis) and the magnetic field, $B$, (along the z-axis), and therefore along the y-axis.

$$E_y = \frac{-\nabla_x T}{S_\phi}\eta B \tag{9.13}$$

Thus, the Nernst signal would be

$$N = \frac{E_y}{-\nabla_x T} = \frac{S_\phi}{\eta}B \tag{9.14}$$

According to this phenomenological equation, the Nernst signal linearly grows with a magnetic field and its slope is proportional to the ratio of transport entropy per vortex to vortex viscosity. But, how can one figure out the magnitude of the latter? One way is to recall that the same viscosity plays a role in flux–flow resistivity. If instead of applying the thermal gradient, one applies a charge current, the vortices would be subject to a Lorentz force (instead of thermal force). Again, it is the viscous force, which leads to a steady-state in which a finite electric field appears along the orientation of the applied current. It is easy to show that the magnitude of the emergent flux–flow resistivity would be

$$\rho = \frac{E_x}{\mathcal{J}_x} = B\phi_0/\eta \tag{9.15}$$

In this equation, $\phi_0 = h/2e$ is the quantum of magnetic flux inside a superconducting vortex. Combining the last two equations, we can conclude that the Nernst signal is given by

$$N = \frac{S_\phi \rho}{\phi_0} \tag{9.16}$$

Using the resistivity and Nernst data in the vortex-liquid regime, data similar to those presented in the two panels of Fig. 9.6 and, thanks to the fact that $\phi_0$ is a universal constant, one can extract the magnitude of $S_\phi$ for a given material. Huebener and his collaborators performed such an analysis in the case of several cuprates near optimal doping [Huebener 1995].

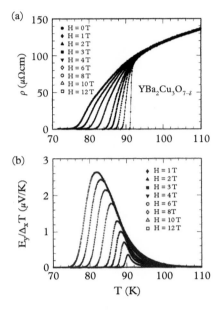

**Figure 9.6** *Nernst signal in the mixed state of a high-temperature superconductor. Reprinted from [Ri et al. 1994], copyright 1994 by the American Physical Society.*

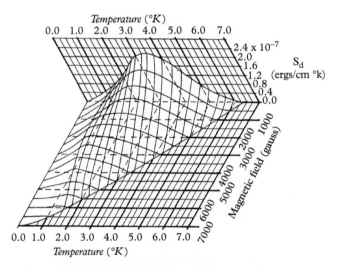

**Figure 9.7** *The transport entropy per vortex as a function of temperature and magnetic field in an indium-lead superconducting alloy. Reprinted from [Solomon and Otter, 1967], copyright 1967 by the American Physical Society.*

What is the physical meaning of the entropy transport per vortex? It is easy to see that $S_\phi$ should vanish both at the critical temperature (where there is no entropy difference between normal and superconducting states) and at zero temperature (where there is no entropy at all). Thus, $S_\phi$ is expected to peak at a temperature between these two limits. It has also a field dependence. When the magnetic field attains the upper critical field, $S_\phi$ should vanish, since there is no difference between the vortex core (where the entropy excess is accumulated) and the surrounding environment. Figure 9.6 represents the transport entropy per vortex as a function of temperature and magnetic field. It is indeed finite at intermediate range of temperature and magnetic field.

A thermodynamic argument, first put forward by Stephen [1968], links the magnitude of the entropy transport per vortex at low magnetic field to the temperature dependence of the lower critical field. If the magnetic field is small enough, vortices are far apart and their energetics can be analysed individually. In this regime, the magnitude of $S_\phi$ can be linked to the magnetic energy stored in a single vortex. In SI units, this energy per unit length of the vortex is

$$u_M = \frac{1}{2} H_{c1} \phi_0 \tag{9.17}$$

This is the energy cost of magnetic expulsion. The entropy associated with this energy is

$$S_\phi = \phi_0 \frac{1}{2} \frac{\partial H_{c1}}{\partial T} \tag{9.18}$$

Huebener uses this simple argument to give an account of low-temperature entropy transport of vortices in niobium. However, it predicts a finite $S_\phi$, even in the vicinity of the critical temperature, in contrast to experimental finding [Huebener 1972]. In general, vortices are not very mobile at low temperature and low magnetic field. Thus, the link between the slope of the lower critical field and the transport entropy suggested by Eq. 9.18 has not been subject to a rigorous experimental verification. A more accessible quantity is the vortex transport entropy in magnetic fields much larger than $H_{c1}$ and intermediate temperatures. But in this case, even a phenomenological treatment in the entire range of temperature and magnetic field is complex [de Lange and Otter 1972].

## 9.3 Superconducting Fluctuations

The first decade of this century saw a revival of interest in the Nernst effect and its link to superconducting fluctuations, following the publication of a *Nature* paper entitled 'Vortex-Like Excitations and the Onset of Superconducting Phase Fluctuations in Underdoped La$_{2-x}$Sr$_x$CuO$_4$' [Xu *et al.* 2000]. The authors reported on Nernst measurements in a cuprate superconductor and how they could shed light on the puzzle of high-temperature superconductivity. An intense debate followed this finding (see Chapter 8) and consensus has been hard to reach. On the other hand, one other consequence of this discovery has been a significant progress in our understanding of the Nernst signal generated by fluctuating superconductivity in the context of a conventional BCS superconductor.

Warming a superconductor above its critical temperature does not wipe out superconductivity. The condensate of Cooper pairs is destroyed at this temperature, but Cooper pairs can survive in limited windows of time and length. Consider, the order parameter with its smooth variation in time and in space. Below $T_c$, in the superconducting state, it has a steady value and does not present any variation with time. The characteristic time scale is thus infinite, because this is the typical time you need to wait before observing a change in the amplitude of the order parameter. Above the critical temperature, the steady-state magnitude of the order parameter is zero. However, it can attain a finite amplitude during a limited time. This finite time scale is basically set by quantum mechanics. Suppose that you have raised the temperature of your superconductor (which we will suppose to be perfectly homogeneous) to $T$, just a little above the well-defined critical temperature, $T_c$. The average thermal energy of the system is now $k_B T$, which is slightly larger than $k_B T_c$. But according to the uncertainty principle, a narrow energy window is concomitant with a wide temporal window. This defines the life expectancy of the Cooper pairs above the critical temperature:

$$\tau_{GL} \sim \frac{\hbar}{k_B(T - T_c)} \tag{9.19}$$

The index specifies the fact that we are referring to the temporal variation of the Ginzburg-Landau order parameter. In other words, during a typical time window of $\tau_{GL}$, you can expect the birth and death of Cooper pairs somewhere in the normal state of your superconductor. This is only an estimation of the order of magnitude. The pre factor is quantified after elaborating a microscopic theory. A second fundamental concept is the superconducting correlation length, $\xi$. This is roughly the typical distance separating two electrons, which can momentarily pair up. In the immediate vicinity of the critical temperature, this distance is very long, which means that electron can find a partner almost anywhere. As the system warms up and superconductivity weakens, fluctuations are confined to small distances and the correlation length shortens. In a dirty superconductor, with the approach of the critical temperature, it increases as

$$\xi = \frac{\xi_0}{\sqrt{\epsilon}} \tag{9.20}$$

The coherence length $\xi_0$ depends on the material. The parameter $\epsilon$ is the so-called reduced temperature:

$$\epsilon = \ln\left(\frac{T}{T_c}\right) \approx \frac{T - T_c}{T_c} \tag{9.21}$$

The standard textbook on fluctuating superconductivity [Larkin and Varlamov 2005] gives an extensive discussion of these Gaussian fluctuations of the superconducting order parameter. Paraconductivity is the best-known physical phenomenon emanating from ephemeral Cooper pairs in the normal state. A metal which approaches the critical temperature from above conducts more than usual because of the opening of a new channel for charge transport due to these short-lived Cooper pairs. This shows itself in the form of resistive transition. Instead of being step like, it becomes rounded in its high-temperature part. Above $T_c$, resistivity deviates downward from its normal-state value before diving to zero. This effect was clearly observed for the first time in amorphous thin films of bismuth [Glover 1967]. It is often called Aslamazov-Larkin fluctuations in reference to the pair of Soviet scientists who, almost simultaneously, elaborated the theory of this effect [Aslamazov and Larkin 1968].

In two dimensions, the theoretical expression for the magnitude of paraconductivity becomes particularly simple. The conductivity per square is expected to be universal (i.e. material independent) and simply linked to the quantum of conductance:

$$\sigma_\square^{sc} = \frac{1}{16\epsilon} \frac{e^2}{\hbar} \qquad (9.22)$$

This should be compared with the conductivity of normal quasi-particles:

$$\sigma_\square^{qp} = \frac{e^2}{\hbar} k_F \ell \qquad (9.23)$$

Thus, paraconductivity can provide a sizeable fraction of conductivity when $\frac{1}{16\epsilon}$ can compete with $k_F \ell$. Such a situation can only happen in dirty systems not very far from the critical temperature. The discovery that the fluctuating superconductivity can be detected by measuring the Nernst signal far above $T_c$ [Pourret 2006] was a genuine surprise. Retrospectively, however, the experimental observation is compatible with the BCS theory and the Boltzmann picture of electronic transport.

Before the discovery of the puzzling Nernst signal in underdoped cuprates [Xu *et al.* 2000], the expression for the thermoelectric response of fluctuating Cooper pairs had not been formulated. Motivated by this discovery, Ussishkin and co-workers proposed such a theory a couple of years later [Ussishkin, Sondhi, and Huse 2002] and derived the following formula for the Nernst response due to Gaussian fluctuations in a two-dimensional superconductor:

$$\alpha_{xy}^{sc} = \frac{1}{6\pi} \frac{k_B e}{\hbar} \frac{\xi^2}{\ell_B^2} \qquad (9.24)$$

In this remarkably simple expression, the only material-dependent parameter is the superconducting correlation length. The relative size of $\xi$ compared to the magnetic length $\ell_B$ sets the magnitude and the temperature dependence of the expected signal, which dies off as a power law.

Also present in the equation is the expression for the quantum of thermoelectric conductance ($k_B e/h = 3.3$ nA K$^{-1}$), seen in the previous chapters. Since $\xi$, the only variable parameter in this equation, can be easily determined for a given superconductor, it is straight forward to check the validity of this expression, provided that one can experimentally measure $\alpha_{xy}$.

As seen in previous chapters, what is directly accessible to the experimentalist is $S_{xy}$ and not $\alpha_{xy}$. In general, the determination of $\alpha_{xy}$ implies a knowledge of diagonal and off-diagonal thermoelectric and electric responses, since

$$S_{xy} = \frac{\sigma_{xx}\alpha_{xy} - \sigma_{xy}\alpha_{xx}}{\sigma_{xx}^2 + \sigma_{xy}^2} \qquad (9.25)$$

But when the Hall angle is very small and the second term is negligible, the expression simplifies to

$$S_{xy} \approx \frac{\alpha_{xy}}{\sigma_{xx}} \qquad (9.26)$$

Thus, in such a particular case, resistivity and the Nernst response suffice to determine $\alpha_{xy}$. Ussishkin and co-workers suggested that Gaussian fluctuations could provide a partial explanation for the Nernst signal observed in the normal state of high-temperature superconductors. However, their comparison with the experimental data on cuprates [Xu 2000] suggested only a qualitative agreement [Ussishkin, Sondhi, and Huse 2002].

In 2006, Nernst measurements were performed on amorphous superconducting thin films of $Nb_{0.85}Si_{0.15}$ [Pourret 2006]. The ground state of this system is a simple superconductor and the normal state is a dirty metal with roughly one electron per atom. Electrons travel about a single atomic distance before being scattered and are uncorrelated among each other. Because of the very low mobility of electrons and the very large amplitude of the Fermi energy, the quasi-particle Nernst signal is vanishingly small. The contribution of short-lived Cooper pairs to the Nernst signal, on the other hand, is not damped by these factors. As a consequence, this system is an ideal place to check for the validity of the theoretical expectations.

Figure 9.8 presents typical Nernst data obtained in these studies [Pourret *et al.* 2006]. Below the critical temperature, a vortex Nernst signal as large as several microvolts per kelvin can be detected (left panel). This signal, as expected, presents a non-monotonous field dependence. It is negligible at low magnetic field and almost vanishes above the upper critical field. It gradually decreases in magnitude as the critical temperature is approached, but does not vanish above $T_c$ as seen in the right panel of the same figure. Interestingly, the signal, which is now orders of magnitude smaller (compare the $y$-axis scales) displays the same non-monotonous profile.

The $\alpha_{xy}$ extracted from Nernst and resistivity data was found to be in excellent agreement with the theoretical expectations of Eq. 9.24. The Fermi energy and mobility (extracted by measuring the Seebeck coefficient and the Hall angle) is such that the estimated contribution of normal quasi-particles becomes three orders of magnitude lower than the measured signal. The results provided compelling confirmation of the

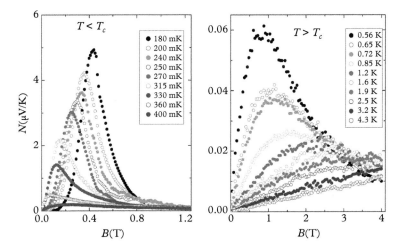

**Figure 9.8** *The Nernst signal as a function of magnetic field for different temperatures in an amorphous superconducting thin film of* $Nb_{0.15}Si_{0.85}$. *Below* $T_c$ *(left), the maximum shifts to lower fields with decreasing temperature. Above* $T_c$ *(right), it shifts in an opposite direction (after [Pourret 2007]).*

theory of Gaussian fluctuations in a conventional BCS superconductor (for a review, see [Pourret *et al.* 2009]).

The validity of Eq. 9.24 is restricted to the limit of $\xi \ll \ell_B$ (i.e. low magnetic fields). When the magnetic length is long enough, the zero-field superconducting correlation length is the only relevant length scale and sets the spatial extension of superconducting fluctuations. This is no more true when the magnetic field increases and the magnetic length shortens, providing an additional source for attenuating fluctuations. This is the fundamental reason for the non-linearity of the field dependence of the Nernst signal in the normal state seen in the date of the right upper panel of Fig. 9.6.

This field scale $H^*$ above the critical temperature mirrors that of the upper critical field below. $H^*$ has been dubbed the ghost critical field [Kapitulnik, Palevski, and Deutscher 1985]. It separates two distinct regimes of fluctuations: below $H^*$, they are governed by $\xi$ in the low-field regime; above, when the magnetic field exceeds the ghost critical field, by $l_B$.

Figure 9.9 [Pourret *et al.* 2007] visualizes the symmetry of the two relevant length scales. The Nernst coefficient measured as a function of temperature and magnetic field is now represented in the $(\ell_B, \xi)$ plane. One of these two length scales is set by temperature, the other by magnetic field. It is straightforward to express the magnetic field as a length $\ell_B^2 = \hbar/2eB$. In the case of temperature, the BCS expression for the coherence length of a dirty superconductor can be employed:

$$\xi_d = \frac{0.36}{\sqrt{\epsilon}} \sqrt{\frac{3\hbar v_F \ell}{2k_B T}} \qquad (9.27)$$

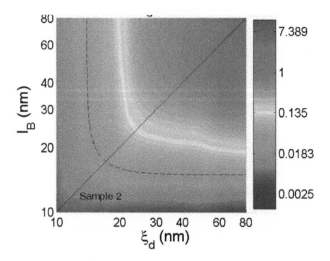

**Figure 9.9** *Logarithmic colour map of the Nernst coefficient in the normal state plotted as a function of two length scales: the magnetic length (extracted from magnetic field) and the superconducting correlation length (extracted from temperature). Note the symmetry of the Nernst coefficient with respect to the diagonal. Reprinted from [Pourret et al. 2007], copyright 2007 by the American Physical Society.*

The figure has a diagonal symmetry respective to the two axes indicating that super-conducting fluctuations are weakened as each of these two length scales, delimiting the spatial extension of fluctuating Cooper pairs, becomes shorter. The diagonal symmetry points to the relevance of all numerical factors employed for the two length scales (including the replacement of $e$ by $2e$ in the expression for the magnetic length. What this figure is telling us is the following: superconducting fluctuations are governed by a single length scale, which becomes rapidly either the magnetic length or the correlation length at the two sides of the border traced by the ghost critical field. This leaves little doubt about the superconducting origin of the measured Nernst signal, even out of the range of validity of Eq. 9.24.

Following the original work by Ussishkin and co-workers, theorists have worked out more refined theories of Gaussian fluctuations in the presence of finite magnetic field [Serbyn *et al.* 2009; Michaeli and Finkelstein 2009]. These theories give a quantitative account of the measured Nernst signal of $Nb_{0.85}Si_{0.15}$ in the entire explored window of temperature and magnetic field.

What is the origin of such a spectacular agreement with theoretical expectations? In this Nernst experiment on a dirty uncorrelated metal, contamination by quasi-particle Nernst signal becomes virtually zero and the measured signal in a wide temperature window above $T_c$ can be attributed to superconducting fluctuations. This is not the case for charge conductivity. Even in a dirty uncorrelated metal, the contribution of superconducting fluctuations to the total conductivity is small compared to the total conductivity. The same is true regarding fluctuating diamagnetism.

In a system as complex and controversial as high-$T_c$ cuprates, the interpretation of the Nernst data has proved to be less straightforward. A recent study [Chang 2010] found that both the magnitude and the temperature dependence of the superconducting Nernst signal in p-doped cuprates is compatible with the expectations of the Gaussian theory. In particular, these authors found that in europium-doped LSCO, the Nernst data are reminiscent of what was found in amorphous thin films of the NbSi alloy.[1]

## 9.4 Relative Magnitude of Normal and Superconducting Nernst Signals

Comparing the relative strength of quasi-particle and short-lived Cooper pairs to the Nernst signal is an instructive exercise. The starting points for such a comparison are Eqs 5.47 and 9.24. One can see that, in both cases, the magnitude of $\alpha_{xy}$ is set by the quantum of thermoelectric conductance times the ratio of a material-dependent length scale to the magnetic length. In three dimensions, the superconducting off-diagonal Peltier conductivity is expected have the following a zero-field slope:

$$\alpha_{xy}^{sc}/B|_{3D} = \frac{1}{6\pi}\frac{k_B e^2}{\hbar^2}\xi \tag{9.28}$$

Compare this to the slope of the superconducting off-diagonal Peltier conductivity for three-dimensional quasi-particles:

$$\alpha_{xy}^{qp}/B|_{3D} = \frac{1}{9\pi}\frac{k_B e^2}{\hbar^2}\frac{k_{dB}^2 \ell^2}{k_F} \tag{9.29}$$

Let us pause for a moment and compare the two. In the case of superconducting fluctuations, the length scale is only set by the superconducting correlation length. In the case of normal quasi-particles, the radius of the Fermi surface, its thermal fuzziness, and the mean-free-path all play a role.

Now, the ratio of these two would be

$$\frac{\alpha_{xy}^{qp}}{\alpha_{xy}^{sc}}|_{3D} = \frac{2}{3}\frac{\ell}{\xi}\frac{k_{dB}^2}{k_F^2}k_F\ell \tag{9.30}$$

Thus, the three dimensions, the clean metallicity ($k_F\ell \gg 1$), and the clean superconductivity ($\ell > \xi_0$) would both favour the domination of the quasi-particle contribution.

---

[1] As this book was in the editing stage, an extensive study of Nernst effect in an n-doped cuprate was reported [Tafti *et al.* 2014]. The study provides compelling evidence for the validity of the standard theory of fluctuations and the relevance of Eq. 9.24 in cuprates over a wide range of doping levels.

This makes the possibility of detecting $\alpha_{xy}^{sc}$ in a clean superconductor hopeless. However, if the $T_c$ happens to be much smaller than the Fermi temperature, near $T_c$ the quasi-particle contribution is damped by $\left(\frac{k_{dB}}{k_F}\right)^2$ and there is a chance.

In two dimensions, the ratio of the two can be written down as

$$\frac{\alpha_{xy}^{qp}}{\alpha_{xy}^{sc}}\bigg|_{2D} = \frac{2}{3}\frac{\ell^2}{\xi^2}\frac{k_{dB}^2}{k_F^2} \tag{9.31}$$

In this case, the $k_F\ell$ term disappears and the dirtiness of the superconductor becomes the central issue. Seen from this perspective, it is not surprising that an experiment on a dirty two-dimensional superconductor with a very low critical temperature to its high Fermi temperature has been particularly successful in detecting the Nernst signal generated by fluctuating Cooper pairs.

# 10

# New Frontiers

This chapter gives a brief account of a few selected topics attracting attention in the last few years. It is fair to expect that they will also be a focus of research during the next few years. Each topic is about thermoelectric response in a particular context. The experimental results are fragmentary and the whole picture is gradually emerging in front of us. The treatment of each of these subjects in this chapter is far from exhaustive. The aim is to give a brief description yielding the particular flavour of the topic to the reader. The interested reader is advised to consult the fast-moving research literature of the subject in question.

## 10.1   Confining Electrons to their Lowest Landau Level

In Chapter 3, we saw that contrary to a thermal gradient, or an electric field, a magnetic field does not affect the Fermi-Dirac distribution of electrons. But, this is only true at low magnetic fields when one can forget quantum mechanical effects. A strong magnetic field introduces a change more drastic than either sharpening or widening the distribution cut-off (what the temperature gradient does) or shifting the whole distribution (what the electric field does). It transforms the continuous spectrum of electrons to a sequence of discrete levels. This phenomenon is called Landau quantization.

Wave-vector happens to be a good quantum number for Bloch electrons. Respecting the Pauli exclusion principle, these electrons will occupy all available states avoiding each other with a finite volume in the $k$-space. The boundary between this occupied volume and what is the left is called the Fermi surface. In an isotropic environment, this would be the surface of a sphere with a radius equal to $k_F$, the Fermi wave-vector. This is not the case of electrons in most metals. The shape of their Fermi surface is set by the symmetry of the underlying lattice and the details of the occupied electronic band.

In the presence of a magnetic field, this Fermi surface is truncated to concentric tubes, often called Landau tubes. The electrons' movement along the magnetic field is not affected by the Lorenz force and the wave-vector along this orientation remains a good quantum number. This is no more true for the movement in the plane perpendicular to the magnetic field. The appropriate quantum number is the Landau level index $n$.

*Fundamentals of Thermoelectricity*. First Edition. Kamran Behnia.
© Kamran Behnia 2015. Published in 2015 by Oxford University Press.

The energy of an electronic state with the quantum numbers, $n$ and $k_z$ is:

$$E(n, k_z) = (n + 1/2)\hbar\omega_c + \frac{\hbar^2 k_\parallel^2}{2m_\parallel} \tag{10.1}$$

The first term of the right-hand side of this equation contains $\omega_c = \frac{eB}{m_\perp}$, the cyclotron frequency. It increases linearly with magnetic field and sets the distance between the quantized energy levels. For electrons in vacuum, the mass is a universal constant, but for those travelling in a solid, the effective mass is set by the band curvature and can substantially differ from the mass of free electrons. Note that both terms in the right side contain the mass, albeit along different orientations, parallel to, or perpendicular to the magnetic field.

Now, as the magnetic field is swept, the first term steadily increases with increasing magnetic field. At any finite magnetic field, there are integers such as $m$, for which:

$$(m + 1/2)\hbar\omega_c > \epsilon_F \tag{10.2}$$

The smallest $m$ defines the lowest empty Landau level. In other words, there are $m$ occupied Landau tubes. This number gradually decreases with increasing magnetic field. Each of these occupied Landau tubes is a one-dimensional Fermi sea of degenerate states. Since all electrons are quenched to Landau tubes and the energy distance between levels is proportional to magnetic field, the degeneracy of each tube is also proportional to magnetic field. As the magnetic field is swept, the Landau tubes increase in diameter and leave the Fermi surface (Fig. 10.1). This leads to quantum oscillations of various measurable physical properties of the system. Such oscillations were indeed discovered in 1930s in the field dependence of resistivity and magnetization. They are called

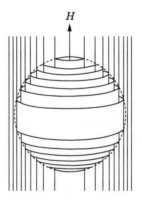

**Figure 10.1** *An illustration of Landau tubes and spherical Fermi surface (after [Shoenberg 1984]).*

Shubnikov-de Haas and de Haas-van Alphen effects, after the name of their discoverers. The standard textbook on this subject is written by Shoenberg [1984].

The period of oscillations is proportional to the extremal section of the Fermi surface perpendicular the magnetic field. Thus, by monitoring the angle dependence of the frequency, one can find the detailed structure of the Fermi surface. Oscillations are detectable only if the Landau quantization is robust enough against smearing by thermal energy and disorder. This means that the experiment is to be done at low enough temperature (to insure $\hbar\omega_c > k_B T$) and high enough magnetic field (to satisfy $\omega_c \tau > 1$). The temperature dependence of the amplitude of oscillations quantifies the ratio of cyclotron to thermal energies and, therefore, yields the magnitude of the cyclotron mass. During the second half of the twentieth century, quantum oscillations have been employed to map the Fermi surface of various metals, and the results are often in good agreement with theoretical calculations based on the band theory [Shoenberg 1984].

The (integer) quantum Hall effect was discovered in 1980s. In a two-dimensional electron gas, quantum oscillations acquire a specific profile. The Hall response becomes quantized and the longitudinal resistivity periodically vanishes. The fundamental reason behind this is that, contrary to the three-dimensional case, electrons have no kinetic energy along the third (i.e. the magnetic field) orientation. Each time the Fermi level lies between two Landau levels, the system is an insulator with no mobile carriers except at its edges. To reach the quantum Hall regime, one needs to attain a magnetic field strong enough to confine all carriers to the lowest Landau levels.

What happens to a three-dimensional system in this situation, when the cyclotron energy becomes comparable to the Fermi energy $\epsilon_F$? There is no well-established answer to this question.

For ordinary bulk metals, the magnetic field necessary to attain this so-called quantum limit is several thousands of Tesla, well beyond the limits of current technology. In a heavy-electron metal, the magnitude of the Fermi energy is significantly reduced because of the heaviness of electrons, but since the same is true for the cyclotron energy, the quantum limit does not become accessible. For a more transparent formulation, one should compare the magnetic length $\ell_B = \sqrt{\hbar/eB}$ with the average inter-electronic distance or the Fermi wavelength (see Table 10.1). The quantum limit is attained when these two length scales become comparable. A field of 10 T corresponds to a magnetic length of $\ell_B \sim 8$ nm, an order of magnitude longer than the typical interatomic distance in solids. To keep only a few Landau levels occupied at a magnetic field of this range, one requires an electronic system much more dilute than copper, which has basically one mobile electron per atom.

In bulk systems, this situation is realized either in doped semiconductors or in semimetals such as bismuth and graphite. In bismuth, the carrier density of holes is $3 \times 10^{17}$ cm$^{-3}$, which means that there is roughly one itinerant electron per $10^5$ atom. Carrier density in graphite is an order of magnitude larger. But, since it is a layered material and its Fermi surface consists of very elongated ellipsoids, the cross section of its Fermi surface is as small as bismuth. In both these systems, when the field is aligned along the high-symmetry axis, a few Landau levels are occupied in presence of a magnetic field as strong as 10 T.

**Table 10.1** *Two distinct limits: above the first threshold field, quantum oscillations become detectable. When the magnetic field exceeds the second threshold filed, electrons are confined to their lowest Landau limit. In semimetals such as bismuth and graphite, this limit is accessible with available magnetic fields.*

|  | Magnetic quantization | Quantum limit |
|---|---|---|
| Time-scale criterion | $\omega_c > \tau^{-1}$ | $\hbar\omega_c > \epsilon_F$ |
| Length-scale criterion | $\ell_B < (\lambda_F \ell_e)^{1/2}$ | $< \ell_B \lambda_F$ |
| Threshold field in copper | $\sim$5 T | $\sim 5 \times 10^4$ T |
| Threshold field in bismuth | $\sim$0.1 T | 9 T |
| Threshold field in graphite | $\sim$0.1 T | 7.5 T |
| Threshold field in cuprates ($p = 0.12$) | 25 T | $\sim$600 T |

The amplitude and profile of quantum oscillations of the thermoelectric response in the vicinity of the quantum limit has become a subject of interest during the last few years. Unexpectedly large oscillations of the Nernst response were reported in bismuth [Behnia 2007]. Later, Nernst oscillations of large amplitude and asymmetric profile were seen in graphite [Zhu 2010]. They are reproduced in Fig. 10.2. In both these systems, the large oscillatory Nernst response (much larger than the monotonous background) emerges only when a few Landau tubes are left. Early studies on both bismuth [Mangez, Issi, and Heremans 1976] and graphite [Woollam 1971] had already reported on quantum oscillations of thermoelectric coefficients, but not where extended to low enough temperatures and/or large enough magnetic fields to detect the spectacular peaks seen in Fig. 10.2.

Subsequent studies led to the observation of giant Nernst oscillations in high-mobility semiconductors such as $Bi_2Se_3$ [Fauqué *et al.* 2013] or $SrTiO_{3-\delta}$ [Lin *et al.* 2013]. These are semiconductors sufficiently doped to be pushed to the metallic side of a metal–insulator transition. They possess a sharp tiny Fermi surface, which gives rise to a large oscillatory thermoelectric response, in particular in the Nernst channel. This seems to be a generic property of bulk systems in the vicinity of the quantum limit as recently observed in the case of a topological crystalline 'insulator' with a bulk Fermi surface [Liang *et al.* 2013].

By scrutinizing old scientific literature, one can find reports from as early as 1959 on oscillations of the Nernst coefficient with a large amplitude and low frequency in metals such as elemental zinc [Bergeron, Grenier, and Reynolds 1959]. Similar large oscillations were also reported in the case of a degenerate semiconductor, iron-doped HgSe [Tieke *et al.* 1996]. Thus, the observation is much older than what was thought at the moment of their rediscovery in late 2000s.

In the case of two-dimensional systems, a number of experiments have been reported and discussed in a review article by Fletcher [1999]. As a consequence of a strong coupling between two-dimensional mobile electrons and the substrate phonons, the phonon

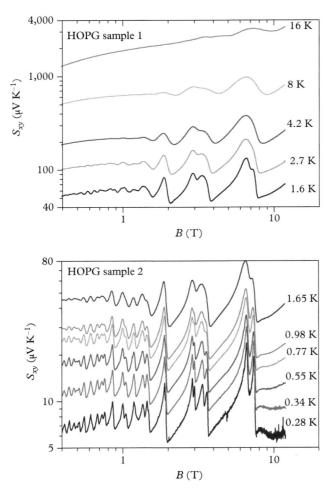

**Figure 10.2** *Quantum oscillations of the Nernst coefficient in graphite. Reprinted by permission from Macmillan Publishers Ltd: Nature [Zhu et al. 2010], copyright 2010.*

drag is large, and diffusive thermoelectricity becomes detectable only at temperatures, as low as 0.5 K [Ruf *et al.* 1988] or even 0.1 K [Ying *et al.* 1994]. Quantum oscillations of thermoelectric coefficients in the quantum Hall regime have been observed not only in high-mobility GaAs/Ga$_{1-x}$Al$_x$As heterojunctions, but also in silicon metal–oxide–semiconductor field-effect transistors [Fletcher *et al.* 1998]. Soon, studies were extended to the fractional quantum Hall regime and the thermoelectricity of composite fermions became a subject of study [Ying *et al.* 1994; Tieke *et al.* 1999].

The discovery of quantum Hall effect in graphene in the middle of the 2000s provided a new two-dimensional electron gas for exploration. Soon afterwards, the thermoelectric response of this particular quantum Hall system was explored [Zuev *et al.* 2009; Wei *et al.* 2009; Checkelsky and Ong 2009]. Figure 10.3 shows the results obtained by Zuev

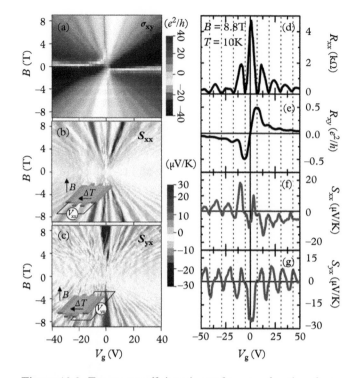

**Figure 10.3** *Transport coefficients in graphene as a function of gate voltage in a fixed magnetic field. Reprinted from [Zuev et al. 2009], copyright 2009 by the American Physical Society. As the carrier concentration is modified, all transport coefficients show anomalies each time a Landau level becomes occupied.*

and co-workers. Remarkably, the phonon-drag contribution in graphene is weak and even at 10 K, the thermoelectric response is essentially diffusive, in contrast to the case of heterojunctions [Tieke *et al.* 1999].

Comparing Figs 10.3 and 10.2, one can see that the profiles of Nernst oscillations in graphene and in graphite are very different. In graphene, each oscillation consists of a pair of negative and positive peaks sandwiching a vanishing signal. In graphite, each oscillation consists of a single asymmetric peak. Empirically, this difference points to a qualitative change in the transverse thermoelectric response in the passage from two dimensions to three.

The two-dimensional case has been the subject of theoretical investigations as early as 1984 [Jonson and Girvin 1984; Oji 1984]. The three-dimensional case attracted attention much later [Bergman and Oganesyan 2010]. It is still unclear if the Mott relation remains relevant across the quantum limit. Theory has successfully reproduced the profiles seen by experiment in both two-dimensional and three-dimensional cases. However, a quantitative description of the experimental thermoelectric response in a particular system with well-established Fermi surface structure has not been achieved yet. Moreover, an experimental investigation of the crossover between single-layer graphene and bulk graphite is also missing.

Nevertheless, in the case of bulk dilute metals, thanks to its sensitivity, the Nernst effect has proved to be a very useful probe of Landau spectrum. In the case of bismuth, angle-resolved Nernst experiments extended to 28 T have mapped the complex Landau spectrum for the whole solid angle (see Fig. 10.4). This complex spectrum has been

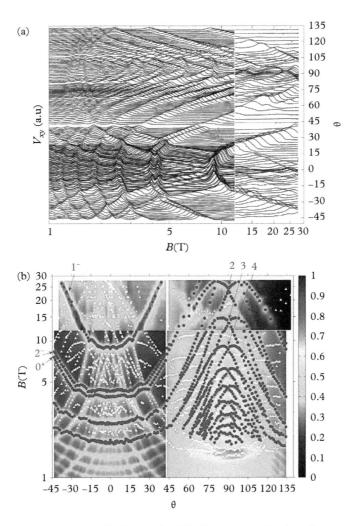

**Figure 10.4** *(a) The Nernst signal in bismuth as a function of magnetic field for different orientations of magnetic field. (b) Nernst peaks superposed on a colour plot of the same data. The data allows us to pin down the angle-resolved Landau spectrum of bismuth, which is surprisingly complex (after [Zhu et al. 2012]).*

found to be in satisfactory agreement with theoretical calculations, pinning down the band parameters of bismuth [Zhu *et al.* 2012].

## 10.2   The Spin Degree of Freedom

The word spintronics, coined to underline the promise of electron's spin to compete with its electric charge as a vehicle of information, is not exactly a household word. But, it has become familiar to the ear of the contemporary condensed-matter physicist. The discovery of the spin Hall effect in the beginning of this millennium [Kato *et al.* 2004] followed its theoretical prediction in the end of the last one [Hirsch 1999] and led to numerous consequences. One can argue that one of them was the discovery of spin seebeck effect a few years later. Contrary to what was believed in the beginning, however, the underlying principles of the two effects are very different. Nevertheless, a new field of research dubbed 'spin caloritronics' [Bauer, Saitoh, and van Wees 2012] or 'thermal spintronics' [Adachi *et al.* 2013] was born and is heading towards maturity.

The underlying cause of the spin Hall effect is the spin–orbit coupling (see Fig. 10.5a). As the electron moves, the magnetic field generated by the motion of such a charged particle can interplay with its spin. A variety of physical effects emerges thanks

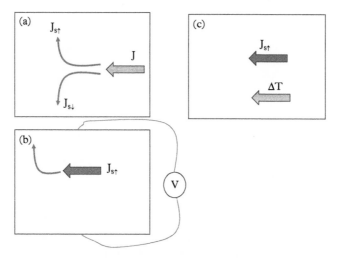

**Figure 10.5** *(a) Spin Hall effect: when a charge current flows in a solid, skew scattering generates an imbalance in the spin of electrons flowing laterally in opposite directions. (b) Inverse spin Hall effect: when a spin current flows in a solid, it will be predominantly scattered along one orientation, giving rise to a finite measurable voltage. (c) Spin Seebeck effect refers to the generation of a spin flow by a temperature gradient.*

to this coupling. In the case of the spin Hall effect, this coupling gives rise to an imbalance between electrons of opposite spins in the direction perpendicular to their flow. The ordinary Hall effect appears in presence of a finite magnetic field and refers to a transverse voltage generated by a longitudinal charge current. In the case of the spin Hall effect, there is no magnetic field and the transverse current is a difference in spin populations. In its simplest formulation, the effect rises because the scattering potential is not insensitive to the spin of the electrons and the skew scattering leads to a preferential orientation of scattered electrons along one lateral direction.

Experimentally, it is not straightforward to directly probe the density of spins. As a consequence, the experimental set-up often used to explore this effect, has been mostly a probe of its inverse, the so-called inverse spin Hall effect (ISHE) (see Fig. 10.5b). In this configuration, spin is injected into a solid, and a transverse voltage caused by the spin flow is detected [Valenzuela and Tinkham 2006]. Spin can be injected to a non-magnetic solid by interfacing it with a ferromagnetic solid. When an electric current is applied across the junction, electrons at the Fermi level of the ferromagnetic metal, which are spin polarized, enter into the non-magnetic metal. Before a spin–flip scattering event changes their spin, they carry their spin along a finite distance. The skew-scattering process expected to cause ISHE should now give rise to a transverse voltage, which changes sign with the sign of injected spin. Since its emergence, the ISHE set-up has become a tool to detect spin currents and was used to detect the spin Seebeck effect.

The spin Seebeck effect refers to the generation of a spin flow by a temperature gradient in a ferromagnet. The spin flow has been detected through the voltage generated by the ISHE in a platinum strip attached to the ferromagnet. The first experiment was on a ferromagnetic metal and it was believed initially that the spin was carried by the conduction electrons. What was surprising, however, was the length scale of several millimetres associated with the experiment. Mobile electrons do not travel a long distance before being scattered in a way to lose their spin identity. This distance is called the spin diffusion length. Roughly, it is the mean-free-path multiplied by a number, which quantifies the rareness of spin-flipping scattering events. It looked puzzling that spin polarization could survive over a length as long as 1 mm.

Following the initial observation, several experiments reported the observation of a spin Seebeck effect in ferromagnetic semiconductors and insulators [Uchida *et al.* 2010; Jaworski *et al.* 2010]. A new interpretation was proposed, according to which carriers of spin are not mobile electrons but magnons, collective excitations of a magnetically ordered solid. Magnons exist independent of the presence or absence of mobile electrons. Like phonons, they can travel along distances much longer than electrons. The typical magnon wavelength increases with decreasing temperature and an extended object like this cannot be easily scattered. This is the fundamental reason for their long mean-free-path and their capacity to become ballistic at low enough temperature.

Because of the sizeable magnitude of its spin Hall coefficient, platinum is widely used to generate and to detect spin flows. A platinum strip can be used to generate a flow of spin-polarized electrons by ISHE. At the interface with the ferromagnet, the spin of these electrons exerts a spin torque on localized magnetic moments inside the insulator and a spin current carried by magnons would be introduced inside the magnet.

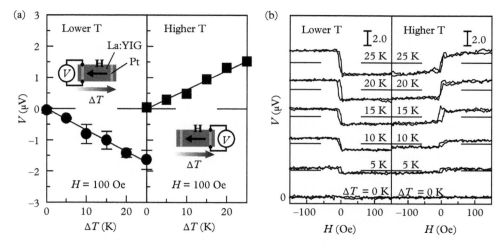

**Figure 10.6** *Spin Seebeck effect in a LaY$_2$Fe$_5$O$_{12}$/Pt junction at room temperature: (a) The dependence of voltage on temperature difference. The finite magnetic field sets the sign of spin polarization. The two sets of data show the measured voltage on two Platinum strips at the cold and the hot ends of the sample. (b) The dependence of the voltage on magnetic field. It changes sign as soon as the spins are flipped by the applied magnetic field. Reprinted from [Adachi* et al. *2013], © IOP Publishing. Reproduced by permission of IOP Publishing. All rights reserved.*

A second Platinum electrode can detect this spin flow through ISHE, due to the spin torque exerted by magnons on spin of conduction electrons at the second interface.

Figure 10.6 shows the set-up and the experimental data obtained in a study on a sample made from an interface between the ferromagnetic insulator LaY$_2$Fe$_5$O$_{12}$ and platinum [Adachi *et al.* 2013]. Two Platinum wires attached to the top surface of single-crystalline LaY$_2$Fe$_5$O$_{12}$, a few micrometres thick and several millimetres long and wide. As seen in the figure, the magnitude of the voltage generated by ISHE is proportional to the applied temperature gradient in both Platinum wires. As expected, its sign is opposite for the warmer and colder ends of the sample. As seen also in the figure, this voltage changes sign as the orientation of magnetic field is reversed.

In this experimental configuration, what is detected is the transverse spin Seebeck effect, in which the direction of the thermal spin injection into the non-magnetic metal is perpendicular to the temperature gradient. It is to be distinguished from the longitudinal spin Seebeck effect, in which the direction of the thermal spin injection is parallel to the temperature gradient. Now, in order to ensure that the voltage is generated by spin flow, one has to flip the spin orientation by reversing the magnetic field and to check an inversion of the voltage sign. But, at least in the case of ferromagnetic metals, such a signal can be contaminated by the Nernst response of the metal. As a consequence, only in the case of insulating ferromagnets, where no Nernst response in the magnetic material is expected, can the longitudinal Seebeck effect be explored.

The spin Seebeck effect is still in its infancy and there is an ongoing debate on the weight of different possible contributions to the observed signal in different materials and configurations. It is believed that in many cases, the non-diffusive component, phonon drag, and magnon drag play an important role. As we saw in Chapter 3, even in the case of ordinary thermoelectricity, the contribution of out-of-equilibrium phonons is yet to be fully understood. The question is open again in the context of spin Seebeck effect,

There is one experimental report on a spin Seebeck effect observed in a non-magnetic semiconductor [Jaworski *et al.* 2012]. The system under study was n-doped InSb, a narrow-gap semiconductor with high-mobility electrons of reduced mass. The system has a very small Fermi surface. Therefore, a magnetic field as small as 2 T confines carriers to their lowest Landau level (see Section 10.1). Electrons confined to this lowest Landau level are all spin polarized, because of the large Zeeman energy of the system. The spin Seebeck effect seen in this configuration is attributed to these spin-polarized electrons, with an eventual phonon-drag component.

There is another class of spin-dependent Seebeck experiments where the main source of the signal is unambiguously provided by the conduction electrons. A recent experiment, for example, reports on a spin current driven by the flow of heat across the interface between a ferromagnet and a non-magnetic metal. In this case, the spin current is due to the spin dependence of the Seebeck coefficient inside the ferromagnetic material. A temperature gradient across an interface between a ferromagnetic and a non-magnetic material drives electrons with polarized spins into the non-magnetic solid [Slachter *et al.* 2010]. These experiments are the thermoelectric analogues of giant magnetoresistance and spin valves.

## 10.3   Nanometric Dimensions

The flow of heat in reduced dimensions attracts growing attention. Atomic and molecular junctions, carbon nanotubes, quantum point contacts, quantum dots, and other devices associated with nanotechnology present interesting electrical and optical properties. Theorists and experiments have been exploring the way thermoelectric transport is modified by drastic reduction in sample dimensions (for a review, see [Dubi and Di Ventra 2011].

Pioneer thermoelectric experiments on quantum point contacts were carried out in the 1990s [Molenkamp *et al.* 1990, 1992]. The results have been the subject of an excellent early experimental review [van Houten *et al.* 1992]. A quantum point contact is a device on top of a two-dimensional electron gas formed in a semiconducting heterostructure (mainly GaAs/AlGaAs). It is a narrow constriction between two electron reservoirs with a width of the order of the electron Fermi wavelength. A gate voltage controls the number of electron waves, which can ballistically travel across the quantum point contact. At low temperatures, the electric conductance across the constriction was found to be quantized as a function of the gate voltage) [van Wees *et al.* 1988]. This was an important experimental result confirming Landuaer's picture of conduction as transmission of electron waves.

The configuration and the results of the first attempt to measure the thermoelectric response of a quantum point contact are shown in Fig. 10.7 [Molenkamp *et al.* 1990].

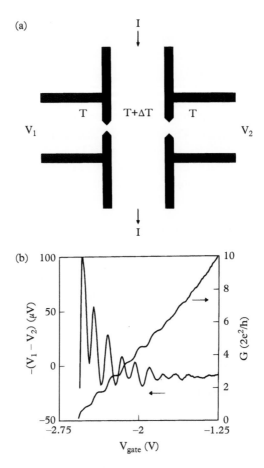

**Figure 10.7** *(a) Schematic representation of the device used to study the thermopower of quantum point contacts. The channel width is 4 μm, with two opposite quantum point contacts at its boundaries. (b) Measured conductance and voltage −(V₁ − V₂) as a function of the gate voltage defining point contact 1. The lattice temperature is kept at 1.65 K and the temperature difference has been generated by applying a current of 5 μA, The gate voltage of the quantum point contact 2 was kept constant. Reprinted from [van Houten et al. 1992], © IOP Publishing. Reproduced by permission of IOP Publishing. All rights reserved.*

There are two point contacts separated by a distance of a few microns. The gate voltage defining point contact 1 is scanned, while that of point contact 2 is kept constant. In this way, any change in the measured signal is caused by a variation in the difference in the two thermoelectric responses. When the point contact resistance exhibits quantized plateaux, strong oscillations in the measured voltage arise, presumably due to the change in the thermopower of quantum point contact 1. The occurrence of such oscillations as well as the hierarchy of the peaks (decreasing in amplitude as the number of conducting channels increases) is what one would expect for the thermoelectric response of a one-dimensional wire with quantized conductance [Lunde and Flensberg 2005]. However, in spite of this similarity, one should be cautious. Rigorously speaking, the Seebeck coefficient of a one-dimensional system is the magnitude of an electric field, produced by and parallel to an applied thermal gradient, divided by the applied temperature. The measured *transverse* thermal voltage is not directly proportional to the Seebeck coefficient response of the quantum point contact.

A second experiment, performed two years later, explored the Peltier and thermal conductance of quantum point contacts [Molenkamp *et al.* 1992]. Here, using a more sophisticated experimental arrangement, a heat flow was generated by a charge flow, which is the Peltier effect. Then, by measuring the thermal gradient created by this heat flow, the thermal conductance of the quantum point contact was studied. As expected, the Peltier coefficient displayed oscillations concomitant with jumps in charge conductance. On the other hand, the thermal conductance displayed plateaus similar to those seen in the case of electrical conductance. The ratio of the jump in thermal and electrical conductances was close to the expectations of the Wiedemann-Franz law. Therefore, the current experimental situation suggests that the Wiedemann-Franz law or the Mott formula remains valid, at least roughly, in the case of one-dimensional electronic transport across a quantum point contact.

There is another topic in mesoscopic thermoelectricity focused on quantum dots [Beenakker and Staring 1992]. In this case, interaction among electrons can play a significant role, through Coulomb blockade. A quantum dot is an electronic system small enough to display quantum mechanical properties as a result of its confinement in all three spatial dimensions. Coulomb blockade occurs at temperatures low enough so that the energy required to charge the quantum dot with one elementary charge is larger than the thermal energy. It leads to oscillations of electric conductance as a function of gate voltage. The period of oscillations is set by the relative ratio of the Fermi energy and the charging energy $e^2/C$, where $e$ is the charge of electron and $C$ is the capacitance of the quantum dot. The oscillations of electric conductance reflect oscillations of the free energy of the dot as a function of the number of electrons present. Theory has predicted [Beenakker and Staring 1992] and experiment has confirmed [Molenkamp *et al.* 1994] that in this situation, the thermoelectric voltage should display 'sawtooth-like oscillations'.

During the following decade, the Seebeck coefficient in carbon nanotubes became the subject of experimental studies [Small 2003; Kong 2005]. This announced the era of research on thermoelectric response in nanometric objects. A new type of experiment, mostly motivated by enhancing the thermoelectric performance, focused on measuring the Seebeck coefficient of nanowires. The list includes thermoelectric studies on nanowires of bismuth [Boukai *et al.* 2006], silicon [Boukai *et al.* 2008; Hochbaum *et al.* 2008], the narrow-gap semiconductor InSb [Seol *et al.* 2007], and the wide-gap semiconductors GaN and ZnO [Lee *et al.* 2009]. In most cases, the magnitude of the Seebeck coefficient is significantly lower than in the bulk system and doping level is not fully controlled. However, the mere fact that the experiments succeed to produce meaningful data constitutes a technological *tour de force*.

Atomic-size metallic contacts provided another interesting playground. In a manner similar to quantum point contacts, the quantization of electric conductance was observed across such junctions. In one spectacular case, jumps in conductance quantized in quantum of conductance could even be observed at room temperature [Rubio, Agraït, and Vieira 1996]. Intriguingly, only a single report on the study of the thermoelectric response in such metallic constrictions can be found [Ludoph and van Ruitenbeek 1999]. The experimental results are shown in Fig. 10.8. A nanometric contact (across metallic gold) was broken by increasing the voltage of an actuator. As seen in the figure, with increasing piezovoltage, the electric conductance diminished from ten quanta of conductance to zero, with plateaus interrupted by jumps. Steps in the Seebeck response

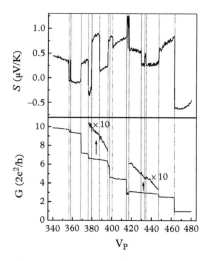

**Figure 10.8** *Steps in conductance G and in thermopower S as the size of metallic junction is reduced by the increase in the voltage $V_P$ of a piezoelectric actuator. The vertical grey lines indicate the corresponding steps in the conductance and thermopower. Reprinted from [Ludoph and van Ruitenbeek 1999], copyright 1999 by the American Physical Society.*

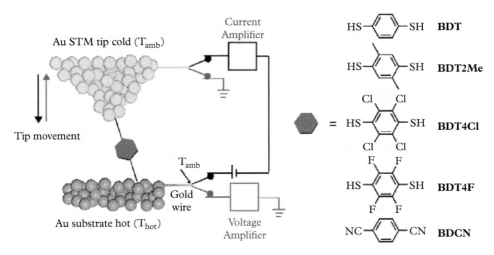

**Figure 10.9** *Set-up for measuring thermoelectric response of a single molecule trapped between a scanning tunnelling microscope (STM) tip and a heated substrate. Reprinted with permission from [Baheti* et al. *2008]. Copyright 2008 American Chemical Society.*

were observed concomitant with these jumps. However, both the sign and the size of these steps were found to be arbitrary. One hopes to see future experiments like this with a better control of the applied temperature gradient and including the measurement of all transport coefficients applied along the broken junction.

Another emerging avenue of research is the thermoelectric study of molecular junctions [Baheti *et al.* 2008]. In one such experiment [Reddy *et al.* 2007], molecules were trapped between two gold electrodes with a temperature difference across them and the Seebeck coefficient of the junction was measured at room temperature (see Fig. 10.9). One of the two electrodes is a heated substrate and the other is a scanning tunnelling microscope tip. The tip approaches the substrate until a threshold conductance is attained. Then, for different values of temperature difference between the substrate and the tip, the voltage difference is measured. The data comprise a distribution of measured voltages, leading to histograms with visible peaks. Thus, the Seebeck coefficient of a 1,4-benzenedithiol molecule was determined to be $8.7 \pm 2.1$ $\mu$V K$^{-1}$. One motivation of such experiments is to obtain information on the electronic spectrum in the energy window between the highest occupied and lowest unoccupied molecular orbits [Baheti *et al.* 2008].

## 10.4   The Link to Information Entropy

Maxwell introduced his demon in 1867. The creature has exerted an everlasting fascination on scientific imagination ever since. It is hardly rivalled by any other ingredient of a thought experiment (Schrödinger's cat is the only serious candidate). The demon has shaken the foundations of the second law of thermodynamics and has been resurrected several times after being declared vanquished.

In the original formulation, the demon is a being capable of monitoring the individual molecules of air contained in two chambers linked by a small aperture. It can open and close the aperture at will. If it does so in a way to allow the faster molecules to enter one chamber and the slower molecules to enter the other, it will create a situation in which one chamber is full of rapid molecules and the other populated by slower ones, leading to different temperatures in the two chambers. By doing so, the demon has created a temperature difference without dispensing energy and thus has violated the second law of thermodynamics. Maxwell's own interpretation was that the second law 'has only a statistical certainty'. However, it was beginning of a long tale, twisting entropy and information (for a brief account, see [Bub 2002]).

In 1929, Szilard made the observation that, to perform its task, the demon needs to know the positions and the velocities of individual molecules and argued that the acquisition of this knowledge would have an entropy cost [Szilard 1929]. The argument that a bit of information costs $k_B ln(2)$ of entropy was elaborated in more detail later by Brillouin [1956]. The second law of thermodynamics was protected from the mischievous creature.

However, decades later, first Landauer [1961] and then Bennett [1973] argued that a reversible measurement by itself does not cost entropy. On the other hand, the result of such a measurement needs to be registered somewhere and it is the stocking of information which costs entropy. Moreover, they identified the specific step at which entropy is produced in the information acquisition process. According to them, somewhat counter-intuitively, it is during the resetting of the memory register that entropy is generated. Thus, what costs entropy is not registering new information, but discarding the old one. These ideas laid the foundations of a field of research focused on the thermodynamics of information processing [Bennett 1982].

It has been argued recently that Landauer's principle, that is, the entropy cost of information erasure, cannot save the second law by itself, unless one accepts it axiomatically [Earman and Norton 1999]. A lively debate has followed [Bennett 2003]. Thus, even after a century and a half, the demon may have not said his last word.

The modern theory of information was founded by Claude Shannon in 1948 [Shannon 1948]. Shanon was the first to recognize that the information contained in a message is proportional to the uncertainty about its content. He quantified this amount of uncertainty by writing the following equation for the quantity he dubbed information entropy for a message written an alphabet $A$:

$$H(A) = -K \sum_{i=1}^{n} p_i log(p_i) \tag{10.3}$$

Here, $K$ is a positive constant defined by the units used. The occurrence probability of the symbol $i$ among the total $n$ symbols of the alphabet $A$ is given by $p_i$, so that one has:

$$\sum_{i=1}^{n} p_i = 1 \tag{10.4}$$

Note that information defined in this way ignores semantics. The equation only focuses on the statistical unpredictability of the transmitted message. Suppose that the alphabet in question has a symbol $j$ so that $p_j = 1$. In such a case, a totally predictable alphabet $H$ is equal to zero and no information is transmitted. According to a celebrated story, Shanon used the word entropy upon von Neumann's suggestion. It was understood, however, that the two concepts, that is, entropy in the information theory and entropy in thermodynamics, were only formal cousins.

In the following decades, it was the turn of the infant field of information theory to fertilize the venerable science of thermodynamics. In 1957, E. T. Jaynes introduced the principle of maximum entropy [Jaynes 1957a, b]. He argued that Shannon's concept can be generalized as a basic element of probability theory and for the construction of prior probabilities based on available evidence. After borrowing the word entropy from the information theory, Jaynes turned his attention to statistical mechanics and to putting the work of Gibbs on a new basis.

In the new approach, maximizing entropy means maximizing the uncertainty in a probability distribution subject to available data (see [Grandy 2003]). Thus, entropy becomes a measure of ignorance or uncertainty of an observer in front of a macroscopic system with many possible configurations of microstates. The amount of uncertainty depends on the information of the observer regarding microstates and, in this sense, it is subjective. However, different observers, with identical amount of information, would build identical probability distributions and would have identical expectations.

As a consequence, and as Jaynes himself put it, 'the entropy of a thermodynamic system is a measure of the degree of ignorance of a person whose sole knowledge about its microstates consists of the values of the macroscopic quantities $X_i$, which define its thermodynamic state. This is a completely "objective" quantity in the sense that it is a function only of the $X_i$, and does not depend on anybody's personality. There is then no reason why it cannot be measured in a laboratory.' [Jaynes 1979].

The thermoelectricity is yet to be inserted in this framework. Since Callen, we know that the Onsager relations provide a thermodynamic basis for the thermoelectric phenomena in which the Seebeck coefficient becomes 'entropy flow per particle' [Callen 1948]. The question which has not been raised, let alone answered, is, what is the amount of uncertainty that the particle carries in its flow? In the closing paragraphs of this book, let us briefly consider the case of a flow electrons in a Fermi sea.

In a free-electron gas, there is a Fermi sea with a depth of $k_F$, the radius of the Fermi sphere. At finite temperature, the surface of this sea is thermally agitated within a thickness of $k_{dB}$, the de Broglie wave-vector defined by Eqs 5.1 and 5.2 (see Fig. 10.10). Now, according to Eq. 5.20, the Seebeck coefficient of this system is

$$S = \frac{\pi}{3} \frac{k_B}{e} \left( \frac{k_{dB}}{k_F} \right)^2 \tag{10.5}$$

Recall that the flowing particles here are identical and indistinguishable fermions. At zero temperature, because of the Pauli exclusion principle, there will only be electrons of opposite spins, with a wave-vector $k = k_F$ which can flow. As the temperature increases,

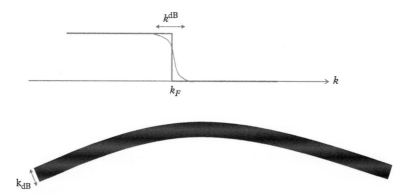

**Figure 10.10** *The Fermi sea is thermally agitated within a thickness of the order of the de Broglie wave-vector, $k_{dB}$.*

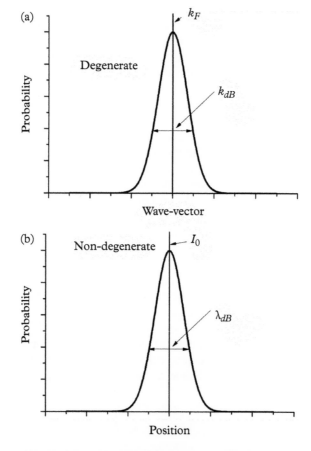

**Figure 10.11** *In a degenerate Fermi system (a), the distribution of the wave-vector of flowing electrons peak at the Fermi wave-vector and has a width of the order of the de Broglie wave-vector. In a non-degenerated Fermi system (b), flowing electrons have a well-defined position with an uncertainty of the de Broglie wavelength.*

the army of flowing electrons consist of those lying within a distance of $k_{dB}$ off the surface. The entropy carried by each electron is a measure of the external observer's ignorance about the identity of the flowing electrons.

The de Broglie thermal wavelength $\lambda_{dB}$ is a widely discussed physical quantity [Yan 2000]. In a system of fermions, it sets the length scale for the quantum to classical passage. The system becomes degenerate when the distance between fermions is smaller than $\lambda_{dB}$. More relevant to our purpose, it can be defined as a measure of the thermodynamic uncertainty in the localization of a particle whose momentum is set by thermal energy [Silvera 1997], an important physical property in a non-degenerate Fermi system. On the other hand, $k_{dB} = \frac{2\pi}{\lambda_{dB}}$ is the thermal thickness of the Fermi-Dirac distribution in a degenerate Fermi system

Thus, in a non-degenerate Fermi system, flowing electrons are particles with thermal momentum. Each of these particles has a position, which is well defined within $\lambda_{dB}$. In the degenerate system, flowing electrons are waves whose wave-vector is well defined to be $k_F$ within an uncertainty of $k_{dB}$. In the former case, the Seebeck coefficient scales with the square of $\lambda_{dB}$ and is thus proportional to the inverse of temperature. In the latter case, it scales with the square of $k_{dB}$ and is linear in temperature. This is illustrated in Fig. 10.11. The uncertainty travelling with each electron is a topic waiting to be explored.

# References

Adachi H., Uchida K., Saitoh E., and Maekawa S. (2013). Theory of the spin Seebeck effect. *Rep. Prog. Phys.* **76**, 036501

van Aken P. B., van Daal H. J., and Buschow K. H. J. (1974). Kondo sideband effects in the Seebeck coefficient of $Ce_{l-x}La_xAl_3$ compounds. *Phys. Lett.* **49A**, 201–203

Akrap A., Ubaldini A., Giannin E., and Forro L. (2014). $Bi_2Te_{3-x}Se_x$ series studied by resistivity and thermopower. *EPL* 107, 57008.

Allgaier R. S. and Scanlon W. W. (1958). Mobility of electrons and holes in PbS, PbSe, and PbTe between room temperature and 4.2°K. *Phys. Rev.* **111**, 1029–1037

Amato A., Jaccard D., Sierro J., Haen P., Lejay P., and Flouquet J. (1989). Transport properties under magnetic fields of the heavy fermion system $CeRu_2Si_2$ and related compounds (Ce,La)$Ru_2Si_2$. *J. Low Temp. Phys.* **77**, 195–208

Ando Y., Miyamoto N., Segawa K., Kawata T., and Terasaki I. (1999). Specific-heat evidence for strong electron correlations in the thermoelectric material (Na,Ca)$Co_2O$. *Phys. Rev. B* **60**, 10580–10583

Ando Y. *et al.* (2000). Carrier concentrations in $Bi_2Sr_{2-z}La_zCuO_{6+\delta}$ single crystals and their relation to the Hall coefficient and thermopower. *Phys. Rev. B* **61**, R14956–R14959

Ando Y. *et al.* (2001). Mobility of the doped holes and the antiferromagnetic correlations in underdoped high-$T_c$ cuprates *Phys. Rev. Lett.* **87**, 017001

Andres K., Graebner J. E., and Ott H. R. (1975). 4f-Virtual-bound-state formation in $CeAl_3$ at low temperatures *Phys. Rev. Lett.* **35**, 1779–1782

Aoki D. and Flouquet J. (2012). Ferromagnetism and superconductivity in uranium compounds. *J. Phys. Soc. Jpn.* **81**, 011003

Aronov A. G., Gal'perin Yu. M., Gurevich V. L., and Kozub V. I. (1981). The Boltzmann-equation description of transport in superconductors. *Adv. Phys.* **30**, 539–592

Asamitsu A., Moritomo Y., and Tokura Y. (1996). Thermoelectric effect in $La_{1-x}Sr_xMnO_3$. *Phys. Rev. B* **53**, R2952–R2955

Ashcroft N. W. and Mermin N. (1976). *Solid State Physics.* Saunders College Publishing, Philadelphia

Aslamazov L. G. and Larkin A. I. (1968). The influence of fluctuation pairing of electrons on the conductivity of normal metal *Phys. Lett. A* **26**, 238–239

Austin I. G. and Mott N. F. (1969). Polarons in crystalline and non-crystalline materials. *Adv. Phys.* **18**, 41–102

Baber W. G. (1937). The contribution to the electrical resistance of metals from collisions between electrons. *Proc. Roy. Soc. A* **158**, 383–396

Baheti K. *et al.* (2008). Probing the chemistry of molecular heterojunctions using thermoelectricity. *Nano Lett.* **8**, 715

Bailyn M. (1958). Transport in metals: Effect of the nonequilibrium phonons, *Phys. Rev.* **112**, 1587

Bailyn M. (1960). Transport in Metals. II. Effect of the phonon spectrum and umklapp processes at high and low temperatures, *Phys. Rev.* **120**, 381404

Bailyn M. (1962). Maximum variational principle for conduction problems in a magnetic field, and the theory of magnon drag, *Phys. Rev.* **126**, 2040–2054

Bailyn M. (1967). Phonon-drag part of the thermoelectric power in metals, *Phys. Rev.* **157**, 480–485

Banerjee A. *et al.* (2008). Transport anomalies across the quantum limit in semimetallic $Bi_{0.96}Sb_{0.04}$. *Phys. Rev. B* **78**, 161103

Bardeen J., Rickayzen G., and Tewordt L. (1959). Theory of the thermal conductivity of superconductors, *Phys. Rev.* **113**, 982–994

Barišić N. *et al.* (2013). Universal quantum oscillations in the underdoped cuprate superconductors. *Nature Phys.* **9**, 761–764

Barnard R. D. (1972). *Thermoelectricity in Metals and Alloys.* Taylor & Francis, London

Barybin A. A. (2008). A Fermi liquid approach to an explanation of the thermoelectric effects experimentally observed in superconductors. *Supercond. Sci. Technol.* **21**, 105005

Bauer G. E. W., Saitoh E., and van Wees B. J. (2012). Spin caloritronics. *Nat. Mater.* **11**, 391–399

Bednorz J. G. and Müller K. A. (1986). Possible high $T_c$ superconductivity in the Ba-La-Cu-O system. *Z. Phys. B* **64**, 189–193

Beenakker C. W. J. and Staring A. A. M. (1992). Theory of the thermopower of a quantum dot. *Phys. Rev. B* **46**, 96679676

Behnia K. (2009). The Nernst effect and the boundaries of the Fermi liquid picture. *J. Phys: Condens. Matter* **21**, 113101

Behnia K., Jaccard D. and Flouquet J. (2004). On the thermoelectricity of correlated electrons in the zero-temperature limit. *J. Phys: Condens. Matter* **16**, 5187

Behnia K., Méasson M. -A., and Kopelevich Y. (2007a). Nernst effect in semimetals: The effective mass and the figure of merit. *Phys. Rev. Lett.* **98**, 076603

Behnia K., Méasson M. -A., and Kopelevich Y. (2007b). Oscillating Nernst-Ettingshausen effect in bismuth across the quantum limit. *Phys. Rev. Lett.* **98**, 166602

Bel R. *et al.* (2004a). Giant Nernst effect in $CeCoIn_5$. *Phys. Rev. Lett.* **92**, 217002

Bel R., Jin H., Behnia K., Flouquet J., and Lejay P. (2004b). Thermoelectricity of $URu_2Si_2$: Giant Nernst effect in the hidden-order state. *Phys. Rev. B* **70**, 220501(R)

Beni G., Kwak J. F., and Chaikin P. M. (1975). Thermoelectric power, Coulomb correlation and charge transfer in TCNQ salts. *Sol. State Commun. B* **17**, 1549–1551

Bennett C. H. (1973). Logical reversibility of computation. *IBM J. Res. Dev.* **17**, 525–532

Bennett C. H. (1982). The thermodynamics of computation–A review. *Int. J. Theoret. Phys.* **21**, 905–940

Bennett C. H. (2003). Notes on Landauers principle, reversible computation, and Maxwell's Demon. *Stud. His. Phil. Mod. Phys.* **34**, 501–510

Bergeron C. J., Grenier C. G., and Reynolds J. M. (1959). Oscillatory Ettinghausen-Nernst effect. *Phys. Rev. Lett.* **2**, 40–41

Bergman D. L. and Oganesyan V. (2010). Theory of dissipationless Nernst effect, *Phys. Rev. Lett.* **104**, 066601

Berggold K., Kriener M., Zobel C., Reichl A., Reuther M., Müller R., Freimuth A., and Lorenz T. (2005). Thermal conductivity, thermopower, and figure of merit of $La_{1-x}Sr_xCoO_3$

Berman R. and Kopp J. (1971). The thermoelectric power of dilute gold-iron alloys. *J. Phys. F: Metal Phys.* **1**, 457–468

Bhargava R. N. (1967). de Haas-van alphen and galvanomagnetic effect in Bi and Bi-Pb alloys. *Phys. Rev.* **156**, 785–797

Bianchi A., Movshovich R., Vekhter I., Pagliuso P. G., and Sarrao J. L. (2003). Avoided antiferromagnetic order and quantum critical point in $CeCoIn_5$. *Phys. Rev. Lett.* **91**, 257001

Bickers N. E., Cox D. L., and Wilkins J. W. (1987). Self-consistent large-N expansion for normal-state properties of dilute magnetic alloys. *Phys. Rev. B* **36**, 2036–2079

Black J., Conwell E. M., Seigle L., and Spencer C.W. (1957). Electrical and optical properties of some $M_2^{v-b}N_3^{vi-b}$ semiconductors. *J. Phys. Chem. Solids* **2**, 240–251

Blatt F. J., Flood D. J., Rowe V., Schroeder P. A., and Cox J. E. (1967). Magnon-drag thermopower in iron, *Phys. Rev. Lett.* **18**, 395–396

Blatt F. J., Garber M., and Scott B. W. (1964). Thermoelectric power of dilute copper alloys. II, *Phys. Rev.* **136**, A729–A737

Blatt F. J. and Kropschot R. H. (1960). Thermoelectric power of dilute copper alloys, *Phys. Rev.* **118**, 480–489

Blatt F. J., Schroeder P. A., Foiles C. L., and Greig D. (1976). *Thermoelectric Power of Metals.* Plenum Press, New York

Boukai A. I., Bunimovich Y., Tahir-Kheli J., Yu J.-K., Goddard III W. A., and Heath J. R. (2008). Silican nanowires as efficient thermoelectric materials. *Nature* **451**, 168–171

Boukai A., Xu K., and Heath J. (2006). Size-dependent transport and thermoelectric properties of individual polycrystalline bismuth nanowires. *Adv. Mat.* **18**, 864

Bourassa R. R., Wang S. Y., and Lengeler B. (1978). Energy dependence of the Fermi surface and thermoelectric power of the noble metals, *Phys. Rev. B* **18**, 1533–1536

Boxus J. and Issi J.-P. (1977). Giant negative phonon-drag thermopower in pure bismuth. *J. Phys. C* **10**, L397–L401

Boxus J., Uher C., Heremans J., and Issi J.-P. (1981). Size dependence of the transport properties of trigonal bismuth. *Phys. Rev. B* **23**, 449–452

Brandt N. B. and Chudinov S. M. (1971). Oscillation effects in semimetallic $Bi_{1-x}Sb_x$ alloys under pressure. *Sov. Phys. JETP* **32**, 815–822

Brandt N. B., Lyubutina L. G., and Kryukova N. A. (1968). Investigation of the electron energy spectrum in Bi-Sb alloys. *Sov. Phys. JETP* **26**, 93–98

Brantut J.-P., Grenier C., Meineke J., Stadler D., Krinner S., Kollath C., Esslinger T., and Georges A. (2014). A thermoelectric heat engine with ultracold atoms, *Science*, 342 713–715

Bridgman P. W. (1924) The connections between the four transverse galvanomagnetic and thermomagnetic phenomena. *Phys. Rev.* **24**, 644–651

Brillouin L. (1956). *Science and Information Theory.* Academic Press, New York.

Bub J. (2001). Maxwell's demon and the thermodynamics of computation. *Stud. His. Phil. Mod. Phys.* **32**, 569–579

Buravov L. I., Fedutin D. N., and Shchegolev F. I. (1971). Mechanism of conductivity of well-conducting complexes on the basis of tetracyanquinodimethyl. *Sov. Phys. JETP* **32**, 612–616

Burke J. R., Houston B., and Savage H. T.(1970). Anisotropy of the Fermi surface of p-Type PbTe. *Phys. Rev. B* **2**, 1977–1988

Butch N. P., Kirshenbaum K., Syers P., Sushkov A. B., Jenkins G. S., Drew H. D., and Paglione J. (2010). Strong surface scattering in ultrahigh-mobility $Bi_2Se_3$ topological insulator crystals. *Phys. Rev. B* **81**, 241301(R)

Butcher P. N. (1990). Thermal and electrical transport formalism for electronic microstructures with many terminals. *J. Phys: Condens. Matter* **2**, 4869–4878

Büttiker M. (1986) Four-terminal phase-coherent conductance. *Phys. Rev. Lett.* **57**, 1761–1764

Cain T. A., Kajdos A. P., and Stemmer S. (2013). La-doped $SrTiO_3$ films with large cryogenic thermoelectric power factors. *Appl. Phys. Lett.* **102**, 182101

Callen H. B. (1945). Thermodynamics and an introduction to Thermostatistics. Wiley, New York

Callen H. B. (1948). The application of Onsager's reciprocal relations to thermoelectric, thermomagnetic and galvanometric effects. *Phys. Rev.* **73**, 1349–1358

Cankurtaran M., Celik H., and Alper T. (1985). Ultrasonic quantum oscillations in semimetallic $Bi_{1-x}Sb_x$ alloys. *J. Phys. F: Met. Phys.* **15**, 391–404.

Cao H., Tian J., Miotkowski I., Shen T., Hu J., Qiao S., and Chen Y. P. (2012). Quantized Hall effect and Shubnikov-de Haas oscillations in highly doped $Bi_2Se_3$: Evidence for layered transport of bulk carriers. *Phys. Rev. Lett.* **108**, 216803

Chai Y. S. *et al.* (2007). Thermopower study of the quasi-one-dimensional organic conductor $(TMTSF)_2PF_6$ SDW state. *Phys. Lett. A* **366**, 513–515

Chaikin P. M. and Beni G. (1976). Thermopower in the correlated hopping regime. *Phys. Rev. B* **13**, 647–651

Chaikin P. M., Kwak J. F., and Epstein A. J. (1979). Evidence for strong coulomb correlations in an organic conductor. *Phys. Rev. Lett.* **42**, 117–182

Chaikin P. M., Kwak J. F., Jones T. E., Garito A. F., and Heege A. J. (1973). Thermoelectric power of tetrathiofulvalinium tetracyanoquinodimethane. *Phys. Rev. Lett.* **31**, 601–604

Chang J. *et al.* (2010). Nernst and Seebeck coefficients of the cuprate superconductor $YBa_2Cu_3O_{6.67}$: A study of Fermi surface reconstruction. *Phys. Rev. Lett.* **104**, 057005

Chang J. *et al.* (2011). Nernst effect in the cuprate superconductor $YBa_2Cu_3O_y$: Broken rotational and translational symmetries. *Phys. Rev. B* **84**, 014507

Chang J. *et al.* (2012). Decrease of upper critical field with underdoping in cuprate superconductors. *Nature Phys.* **8**, 751–756

Chaussy J., Gandit Ph., Matho K., and Ravex A. (1982). Absolute thermoelectric power of AuFe alloys between 0.01 K and 7 K. *J. Low Temp. Phys.* **49**, 167–175

Checkelsky J. G. and Ong N. P. (2009). The thermopower and Nernst effect in graphene in a magnetic field. *Phys. Rev. B* **80**, 081413

Choi E. S., Brooks J. S., Kang H., Jo Y. J., and Kang W. (2005). Resonant Nernst effect in the metallic and field-induced spin density wave states of $(TMTSF)_2ClO_4$. *Phys. Rev. Lett.* **95**, 187001

Choi E., Brooks J., and Qualls J. (2002). Magnetothermopower study of the quasi-two-dimensional organic conductor $\alpha$-$(BEDT$-$TTF)_2KHg(SCN)_4$. *Phys. Rev. B* **65**, 205119

Conwell E. M. (1978). Thermoelectric power of 1:2 tetracyanoquinodimethanide (TCNQ) salts. *Phys. Rev. B* **18**, 1818–1823

Cooper J.R. and Loram J.W. (1996). Some correlations between the thermodynamic and transport properties of high $T_c$ oxides in the normal state. *J. Phys. I* **6**, 2237–2263

Costi T. A. and Hewson A. C. (1993). Transport coefficients of the Anderson model. *J. Phys: Condens. Matter* **5**, L361–L368

Crisp R. S. and Rungis J. (1970). Thermoelectric power and thermal conductivity in the silver-gold alloy system from 3-300 K, *Phil. Mag.* **22**, 217–236

Daou R. *et al.* (2010). Broken rotational symmetry in the pseudogap phase of a high-$T_c$ superconductor. *Nature* **463**, 519–522

Delaire O. *et al.* (2011). Giant anharmonic phonon scattering in PbTe. *Nat. Mater.* **10**, 614–619

Delves R. T. (1965). Thermomagnetic effects in semiconductors and semimetals. *Rep. Prog. Phys.* **28**, 249–289

Dimmock J. O., Melngailis I., and Strauss A. J. (1966). Band structure and laser action in $Pb_xSn_{1-x}Te$. *Phys. Rev. Lett.* **16**, 1193–1196

Doiron-Leyraud N. *et al.* (2007). Quantum oscillations and the Fermi surface in an underdoped high-$T_c$ superconductor. *Nature* **447**, 565–568

Doiron-Leyraud N. *et al.* (2013). Hall, Seebeck, and Nernst coefficients of underdoped $HgBa_2CuO_{4+\delta}$: Fermi-surface reconstruction in an archetypal cuprate superconductor. *Phys. Rev. X* **3**, 021019

Donaghy J. J. and Stewart A. T. (1967). Fermi surface of lithium by positron annihilation. *Phys. Rev.* **164**, 396–398

Dubi Y. and Di Ventra M. (2011). Heat flow and thermoelectricity in atomic and molecular junctions. *Rev. Mod. Phys.* **83**, 131

Dugdale J. S. and Bailyn M. (1967). Anisotropy of relaxation times and phonon drag in the noble metals. *Phys. Rev.* **157**, 485–490

Dumont Y., Ayache C., and Collin G. (2000). Dragging excitation characteristics from thermo-electric power in $Bi_2(Sr_{2-y}La_y) CuO_{6+\delta}$ single crystals. *Phys. Rev. B* **62**, 622–625

Earman J. and Norton J. D. (1999). Exorcist XIV: The wrath of Maxwellšs demon. Part II. From Szilard to Landauer and beyond. *Stud. His. Phil. Mod. Phys.* **30**, 1–40

Edwards P. and Sienko M. J. (1978). Universality aspects of the metal-nonmetal transition in condensed media. *Phys. Rev. B* **17**, 2575–2581

Elizarova M. V. and Gasumyants V. E. (2000). Band spectrum transformation and $T_c$ variation in the $La_{2-x}Sr_xCuO_y$ system in the underdoped and overdoped regime. *Phys. Rev. B* **62**, 5989–5996

Emery V. J. and Kivelson S. A. (1994). Importance of phase fluctuations in superconductors with small superfluid density. *Nature* **374**, 434–437

Fauqué B. *et al.* (2013). Magnetothermoelectric properties of $Bi_2Se_3$. *Phys. Rev. B* **87**, 035133

Fischer K. (1967). Self-consistent treatment of the Kondo effect. *Phys. Rev.* **158**, 613–622

Fletcher R. (1999). Magnetothermoelectric effects in semiconductor systems, *Semicond. Sci. Technol.* **14**, R1–R15

Fletcher R., Pudalov V. M., and Cao S. (1998). Diffusion thermopower of a silicon inversion layer at low magnetic fields. *Phys. Rev. B* **57**, 7174–7181

Flouquet J. (2005). On the heavy fermion road. *Prog. Low Temp. Phys.* **15**, 139–281

Foo M. L., Wang Y., Watauchi S., Zandbergen H. W., He T., Cava R. J., and Ong N. P. (2004). Charge ordering, commensurability, and metallicity in the phase diagram of the layered $Na_xCoO_2$. *Phys. Rev. Lett.* **92**, 247001

Frederikse H. P. R. (1953). Thermoelectric power of germanium below room temperature. *Phys. Rev.* **92**, 248

Frederikse H. P. R., Thurber W. R., and Hosler W. R. (1964). Electronic transport in strontium titanate. *Phys. Rev.* **134**, A442–A445

Fritzsch H. (1971). General expression for thermoelectric power. *Solid State Commun.* **9** 1813–1815

Fu L. and Kane C. L. (2007). Topological insulators with inversion symmetry. *Phys. Rev. B* **76**, 045302

Fujita K., Mochida T., and Nakamura K. (2001). High-temperature thermoelectric properties of $Na_xCo_2$-d single crystals. *Jpn. J. Appl. Phys.* **40**, 4644–4647

Fulkerson W., Moore J. P., Williams R. K., Graves R. S., and McElroy D. L. (1968). Thermal conductivity, electrical resistivity, and Seebeck coefficient of silicon from 100 to 1300K. *Phys. Rev.* **167**, 765–782

Funahashi R. and Shikano M. (2002). $Bi_2Sr_2Co_2O_y$ whiskers with high thermoelectric figure of merit. *Appl. Phys. Lett.* **81**, 1459

Gallo C. F., Chandrasekhar B. S., and Sutter P. H. (1963). Transport Properties of Bismuth Single Crystals. *J. Appl. Phys.* **34**, 144

Galperin Y. M., Gurevich V. L., and Kozub V. I. (1974). Thermoelectric effects in superconduct-ors. *J. Exp. Theor. Phys.* **66**, 1387–1397

Galperin Y. M., Gurevich V. L., Kozub V. I., and Shelankov A. L. (2002). Theory of thermoelectric phenomena in superconductors. *Phys. Rev. B* **65**, 064531

Geballe T. H. and Hull G. W. (1954). Seebeck effect in germanium, *Phys. Rev.* **94**, 1134–1140

Geballe T. H. and Hull G. W. (1955). Seebeck effect in silicon, *Phys. Rev.* **98**, 940–947

Gegenwart P., Si Q., and Steglich F. (2008). Quantum criticality in heavy-fermion metals. *Nat. Phys.* **4**, 186–197

Giazotto F., Heikkilä T. T., Luukanen A., Savin A. M., and Pekola J. P. (2006). Opportunities for mesoscopics in thermometry and refrigeration: Physics and applications. *Rev. Mod. Phys.* **78**, 217–274

Ginzburg V. L. (1991). Thermoelectric effects in the superconducting state, *Sov. Phys.Usp.* **34**, 101–107

Glover R. E. (1967). Ideal resistive transition of a superconductor. *Phys. Lett. A* **25**, 542–544

Gold A. V., MacDonald D. K. C., Pearson W. B., and Templeton I. M. (1960). The thermopower of pure copper, *Phil. Mag.* **5**, 765–786

Goldsmid H. J. (2010). *Introduction to Thermoelectricity*. Springer-Verlag, Berlin and Heidelberg

Goldsmid H. J. and Douglas R.W. (1954). The use of semiconductors in thermoelectric refrigeration. *Br. J. Appl. Phys.* **5**, 386–390

Grandy T. W. Jr. (2008). Entropy and the time evolution of macroscopic systems. Oxford University Press, Oxford

Guénault A. M. (1971). A Physical picture for phonon drag thermoelectric power. *J. Phys. F: Metal Phys.* **1**, 373–376

Guénault A. M. and Hawksworth D. G. (1977). Thermoelectric power of the pure noble metals at low temperatures. *J. Phys. F: Met. Phys.* **7**, L219–L222

Gurevich Yu. G. and Mashkevich O. L. (1989). The electron-phonon drag and transport phenomena in semiconductors. *Physics Reports* **181**, 327–394

de Haas W. J. and van den Berg G. J. (1936). The electrical resistance of gold and silver at low temperatures. *Physica* **3**, 440–449

Ham F. S. (1962). Energy bands of alkali metals. II. Fermi surface, *Phys. Rev.* **128**, 2524–2541

Hanna I. I. and Sondheimer E. H. (1957). Electric and lattice conduction in metals, *Proc. Roy. Soc. A* **239**, 247–266

Hansen O. P., Cheruvier E., Michenaud J.-P., and Issi J.-P. (1978). Diffusion thermoelectric power of bismuth in a magnetic field. *J. Phys. C: Solid State Phys.* **11**, 1825–1840

Van Harlingen D. J. (1982). Thermoelectric effects in the superconducting state. *Physica B* **109 & 110**, 1710–1721

Van Harlingen D. J., Heidel D. F., and Garland J. C. (1980). Experimental study of thermoelectricity in superconducting indium. *Phys. Rev. B* **21**, 1842–1857

Harman T. C., Honig J. M., Fischler S., Paladino A. E., and Button M. J. (1964). Oriented single-crystal bismuth Nernst-Ettingshausen refrigerators. *Appl. Phys. Lett.* **4**, 77–79

Hartmann S., Oeschler N., Krellner C., Geibel C., Paschen S., and Steglich F. (2010). Thermopower evidence for an abrupt Fermi surface change at the quantum critical point of YbRh$_2$Si$_2$. *Phys. Rev. Lett.* **104**, 096401

Hasan M. Z. and Kane C. L. (2010). Colloquium: Topological insulators. *Rev. Mod. Phys.* **82**, 3045

Hassinger E., Knebel G., Matsuda T. D., Aoki D., Taufour V., and Flouquet J. (2010). Similarity of the Fermi surface in the hidden order state and in the antiferromagnetic state of URu$_2$Si$_2$. *Phys. Rev. Lett.* **105**, 216409

Heikes R. R., Miller R. C., and Mazelsky R. (1964). Magnetic and electrical anomalies in LaCoO$_3$. *Physica* **30**, 1600–1608

Heikes R. R. and Ure, Jr. R. W. (1961). *Thermoelectricity: Science and Engineering*. Interscience Publishers, New York, London

Heremans J. and Hansen O. P. (1979). Influence of non-parabolicity on intervalley electron-phonon scattering; the case of bismuth. *J. Phys. C: Solid State Phys.* **11**, 1825–1840

Heremans J. P., Jovovic V., Toberer E. S., Saramat A., Kurosaki K., Charoenphakdee A., Yamanaka S., and Snyder G. J.(2008). Enhancement of thermoelectric efficiency in PbTe by distortion of the electronic density of states. *Science* **321**, 554–557

Herring C. (1954). Theory of the thermoelectric power of semiconductors, *Phys. Rev.* **96**, 1163–1187

Herring C., Geballe T. H., and Kunzler J. E. (1958). Phonon-drag thermomagnetic effects in n-type germanium. I. general survey, *Phys. Rev.* **111**, 36–57

Hirsch J. E. (1999). Spin Hall effect. *Phys. Rev. Lett.* **83**, 1834–1837

Hiruma K. and Miura N. (1983). Magnetoresistance study of Bi and Bi-Sb alloys in high magnetic fields. II. Landau levels and semimetal-semiconductor transition. *J. Phys. Soc. Jpn.* **52**, 2118–2127

Hochbaum A. I., Chen R., Delgado R. D., Liang W., Garnett E. C., Najarian M., Majumdar A., and P. Yang P. (2008). Enhanced thermoelectric performance of rough silicon nanowires. *Nature* **451**, 163–167

van Houten H., Molenkamp L. W., Beenakker C. W. J., and Foxon C. T (1992). Thermoelectric properties of quantum point contacts. *Semicond. Sci. Technol.* 7, B215–B221

Huebener R. P. (1966). Effect of phonon drag on the electrical resistivity of metals. *Phys. Rev.* **146**, 502–505

Huebener R. P. (1972). Thermoelectricity in metals and alloys. *Sol. State Phys.* **27**, 63–134

Huebener R. P. (1979). *Magnetic Flux Structures in Superconductors*. Springer, Heidelberg

Huebener R. P. (1995). Superconductors in a temperature gradient. *Supercond. Sci. Technol.* 8 189–198

Hussey N. E. (2005). Non-generality of the Kadowaki–Woods Ratio in Correlated Oxides. *J. Phys. Soc. Jpn.* **74**, 1107–1111

Imada M., Fujimori A., and Tokura Y. (1998). Metal-insulator transitions. *Rev. Mod. Phys.* **70**, 1039–1263

Imry Y. and Landauer R. (1999). Conductance viewed as transmission. *Rev. Mod. Phys.* **71**, S306–S312

Ino A. *et al.* (2002). Doping-dependent evolution of the electronic structure of $La_{2-x}Sr_xCuO_4$ in the superconducting and metallic phases. *Phys. Rev. B* **65**, 094504

Issi J.-P. (1979). Low temperature transport properties of the group V semimetals. *Aust. J. Phys.* **32**, 585–628

Issi J.-P. and Mangez J. H. (1972). Size dependence of the transport properties of bismuth in the phonon-drag region. *Phys. Rev. B* **6**, 4429–4431

Iwasaki K., Ito T., Nagasaki T., Arita Y., Yoshino M., and Matsui T. (2008). Thermoelectric properties of polycrystalline $La_{1-x}Sr_xCoO_3$. *J. Solid State Chem.* **181**, 3145–3150

Izawa K., Behnia K., Matsuda Y., Shishido H., Settai R., Onuki Y., and Flouquet J. (2007). Thermoelectric response near a quantum critical point: The case of $CeCoIn_5$. *Phys. Rev. Lett.* **99**, 147005

Jaccard D., Behnia K., Sierro J., and Flouquet J. (1992). Transport measurements of heavy fermion superconductors. *Phys. Scr.* **T45**, 130–134

Jaccard D., Mignot J. M., Bellarbi B., Benoît A., Braun H. F., and Sierro J. (1985). High pressure transport coefficients of the heavy fermion superconductor $CeCu_2Si_2$. *J. Magn. Magn. Mat.* **47-48**, 23–29.

Jaccard D. and Sierro J. (1982). Thermoelectric power of some intermediate valence compounds. P. Wachter and H. Boppart (eds). *Valence Instabilities*. North Holland, Amsterdam, pp. 409–413

Jacobson D. M. and Ertl M. E. (1972). Phonon drag thermomagnetism in bismuth at low temperatures. *J. Phys. D: Appl. Phys.* **5**, 1358–1366

Jaime M., Lin P., Salamon M. B., and Han P. D. (1998). Low-temperature electrical transport and double exchange in La0.67(Pb,Ca)?0.33MnO3. *Phys. Rev. B* **58**, R5901–R5904

Jain A. L.(1959). Temperature dependence of the electrical properties of bismuth-antimony alloys. *Phys. Rev.* **114**, 1518–1528

Jandl P. and Birkholz U. (1994). Thermogalvanomagnetic properties of Sn-doped $Bi_{95}Sb_5$ and its application for solid state cooling. *J. Appl. Phys.* **76**, 7351

Jaworski C. M., Myers R. C., Johnston-Halperin E., and Heremans J. P. (2012). Giant spin Seebeck effect in a non-magnetic material. *Nature* **487**, 210–213

Jaworski C. M., Yang J., Mack S., Awschalom D. D., Heremans J. P., and Myers R. C. (2010). Observation of the spin-Seebeck effect in a ferromagnetic semiconductor. *Nat. Mat.* **9**, 898–903

Jaynes E. T. (1957a). Information theory and statistical mechanics. *Phys. Rev.* **106**, 620–630

Jaynes E. T. (1957b). Information theory and statistical mechanics.II *Phys. Rev.* **108**, 171–190

Jaynes E. T. (1979) Where do we stand on maximum entropy. In R. D. Levine and M. Tribus (eds). *The Maximum Entropy Formalism*, MIT Press. Cambridge

Jensen J. D., Houston B., and Burke J. R. (1978). Fermi-surface parameters of p-type PbTe as a function of carrier density. *Phys. Rev. B* **18**, 5567–5572

Jeong C., Kim R., Luisier M., Datta S., and Lundstrom M. (2010). On Landauer versus Boltzmann and full band versus effective mass evaluation of thermoelectric transport coefficients. *J. Appl. Phys.* **107**, 023707

Jérome D. (2012). Organic superconductors: When correlations and magnetism walk in. *J. Supercond. Nov. Magn.* **25**, 633–655

Jérome D., Mazaud A., Ribault M., and Bechgaard K. (1980). Superconductivity in a synthetic organic conductor $(TMTSF)_2PF_6$. *J. Physique Lett.* **41**, 95–98

Jones H. and Zener C. (1934). The theory of the change is resistance in a magnetic field. *Proc. Royal Soc. London A* **145**, 264–277

Jonson M. and Girvin S. M. (1984). Thermoelectric effect in a weakly disordered inversion layer subject to a quantizing magnetic field. *Phys. Rev. B* **29**, 1939–1946

Joynt R. and Taillefer L. (2002). The superconducting phases of $UPt_3$. *Rev. Mod. Phys.* **74**, 235–294

Kaden E. and Günter H.-L. (1984). The thermoelectric power of P-Ge at low temperatures. *Phys. Stat. Sol. B* **126**, 733–740

Kadowaki K. and Woods S. B. (1986). Universal relationship of the resistivity and specific heat in heavy-fermion compounds. *Sol. State Commun.*, **58**, 507–509

Kang W., Hannahs S. T., Chiang L. Y., Upasani R., and P. M. Chaikin P. M.(1992). Magnetothermopower study of $(TMTSF)_2PF_6$ (where TMTSF is tetramethyltetraselenafulvalene). *Phys. Rev. B* **45**, 13566–13571

Kang W., Osada T., Jo Y. J., and Kang H. (2007). Interlayer magnetoresistance of quasi-one-dimensional layered organic conductors. *Phys. Rev. Lett.* **45**, 017002

Kapitulnik A., Palevski A., and Deutscher G. (1985). Inhomogeneity effects on the magneto-resistance and the ghost critical field above Tc in thin mixture films of In-Ge. *J. Phys. C* **18**, 1305–1312

Kato Y. K., Myers R. C., Gossard A. C., and Awschalom D. D. (2004). Observation of the spin hall effect in semiconductors. *Science* **306**, 1910–1913

Kawata T., Iguchi Y., Itoh T., Takahata K., and Terasaki I. (1999). Na-site substitution effects on the thermoelectric properties of $NaCo_2O_4$. *Phys. Rev. B* **60**, 10584–10587

Keyes R. W. (1959). High-temperature thermal conductivity of insulating crystals: relationship to the melting point. *Phys. Rev.* **115**, 564–567

Kilic K. and Celik H. (1994). Quantum oscillations and Fermi surfaces of Sn- and Pb-doped Bi. *J. Phys: Condens. Matter* **6**, 3707–3718

Köhler H. (1976a). Non-parabolic E(k) relation of the lowest conduction band in $Bi_2Te_3$. *Physica Status Solidi (B)* **73**, 95–104

Köhler H. (1976b). Non-parabolicity of the highest valence band of $Bi_2Te_3$ from Shubnikov-de Haas effect. *Physica Status Solidi (B)* **74**, 591–600

Köhler H. and Wöchner E. (1975). The g-factor of the conduction electrons in $Bi_2Se_3$. *Physica Status Solidi (B)* **67**, 665–675

Kokalj J., Hussey N. E., and McKenzie R. H. (2012). Transport properties of the metallic state of overdoped cuprate superconductors from an anisotropic marginal Fermi liquid model. *Phys. Rev. B* **86**, 045132

Koláček J. and Lipavský P. (2005). Thermopower in superconductors: Temperature dependence of magnetic flux in a bimetallic loop. *Phys. Rev. B* **71**, 092503

Kondo J. (1964). Resistance minimum in dilute magnetic alloys. *Prog. Theoret. Phys.* **32**, 37–49

Kondo J. (1965). Giant thermo-electric power of dilute magnetic alloys. *Prog. Theoret. Phys.* **34**, 372–382

Kondo S. *et al.* (1997). $LiV_2O_4$: A heavy fermion transition metal oxide. *Phys. Rev. Lett.* **78**, 3729–3732

Kong W. J., Lu L., Zhu H. W., Wei B. Q., and Wu D. H. (2005). Thermoelectric power of a single-walled carbon nanotubes strand. *J. Phys. Condens. Matter* **17**, 1923–1928

Konstantinović Z. *et al.* (2002). Thermopower in the strongly overdoped region of single-layer $Bi_2Sr_2CuO_{6\delta}$ superconductor *Phys. Rev. B* 66, 020503(R)

Kooi C. F., Horst R. B., and Cuff K. F. (1968). Thermoelectric-thermomagnetic energy converter staging. *J. Appl. Phys.* **39**, 4257–4263

Kopp J. (1975). Single impurity Kondo effect in gold: I. Thermopower. *J. Phys. F: Met. Phys.* **5**, 1211–1216

Korenblit I. Ya., Kusnetsov M. E., and Shalyt S. S. (1969). Thermal EMF and thermomagnetic properties of bismuth at low temperature *Sov. Phys. JETP* **29**, 4–10

Lahoud E. *et al.* (2013). Evolution of the Fermi surface of a doped topological insulator with carrier concentration. *Phys. Rev. B* **88**, 195107

Lakner M. and v. Löhneysen H. (1993). Thermoelectric power of a disordered metal near the metal-insulator transition. *Phys. Rev. Lett.* **70**, 3475–3478

Laliberte F. *et al.* (2011). Fermi-surface reconstruction by stripe order in cuprate superconductors. *Nat. Commun.* **2**, 432

LaLonde A. D., Pei Y., Wang H., and Snyder G. J. (2011). Lead telluride alloy thermoelectrics. *Mat. Today*, **14**, 526–532

Landauer R. (1957). Spatial variation of currents and fields due to localized scatterers in metallic conduction. *IBM J. Res. Dev.* **1**, 223–231

Landauer R. (1961). Dissipation and heat generation in the computing process. *IBM J. Res. Dev.* **5**, 183–191

De Lange O. L. and Otter Jr. F. A. (1972). Low-field entropy of vortices in superconductors. *J. Phys. Chem. Solids* **33**, 1571–1581

Larkin A. I. and Varlamov A. A. (2005). *Theory of Fluctuations in Superconductors*. Oxford University Press, Oxford.

Laughlin R. B. (2014). Fermi-liquid computation of the phase diagram of high-$T_c$ cuprate superconductors with an orbital antiferromagnetic pseudogap. *Phys. Rev. Lett.* **112**, 017004

Lebed A. G. (1986). Anisotropy of an instability for a spin density wave induced by a magnetic field in a Q1D conductor. *JETP Lett.* **43**, 174–177

LeBoeuf D. *et al.* (2007). Electron pockets in the Fermi surface of hole-doped high-$T_c$ superconductors, *Nature* **450**, 533–536

Lee P. A., Nagaosa N., and Wen X.- G. (2006). Doping a Mott insulator: Physics of high-temperature superconductivity. *Rev. Mod. Phys.* **78**, 17

Lee C.-H., Yi G.-C., Zuev Y. M., and Kim P. (2009). Thermoelectric power measurements of wide band gap semiconducting nanowires.*Appl. Phys. Lett.* **94**, 022106

Lenoir B., Cassax M., Michenaud J.-P., Scherrer H., and Scherrer S. (1996). Transport properties of Bi-rich Bi-Sb alloys. *J. Phys. Chem. Solids* **57**, 89–99

Liang T., Gibson Q., Xiong J., Hirschberger M., Koduvayur S. P., Cava R. J., and Ong N. P. (2013). Evidence for massive bulk Dirac fermions in Pb1-xSnxSe from Nernst and thermopower experiments. *Nat. Commun.* **4**, 2696

Lifshitz I. M. (1960). Anomalies of electron characteristics of a metal in the high pressure region. *Sov. Phys. JETP* **11**, 1130–1135

Limelette P., Hébert S., Hardy V., Frésard R., Simon Ch., and Maignan A. (2006). Scaling behavior in thermoelectric misfit cobalt oxides. *Phys. Rev. Lett.* **97**, 046601

Lin X., Zhu Z., Fauqué B., and Behnia K. (2013). Fermi surface of the most dilute superconductor. *Phys. Rev. X* **3**, 021002

Link P., Jaccard D., and Lejay P. (1996). The thermoelectric power of $CePd_2Si_2$ and $CeCu_2Ge_2$ at very high pressure. *Physica B*, **225**, 207–213

Littlewood P. B. *et al.* (2010). Band structure of SnTe studied by photoemission spectroscopy. *Phys. Rev. Lett.* **105**, 086404

Liu Y. and Allen R. E. (1995). Electronic structure of the semimetals Bi and Sb. *Phys. Rev. B* **52**, 1556–1577

Liu X., Sidorenko A., Wagner S., Ziegler P., and Löhneysen H. v. (1996). Electronic transport processes in heavily doped uncompensated and compensated silicon as probed by the thermoelectric power. *Phys. Rev. Lett.* **77**, 3395–3398

Logvenov G. Yu., Kartsovnik M. V., Ito H., and Ishiguro T. (1997). Seebeck and Nernst effects in the mixed state of the two-band organic superconductors κ-(BEDT-TTF)$_2$Cu(NCS)$_2$ and κ-(BEDT-TTF)$_2$Cu[N(CN)$_2$]Br. *Synth. Metals* **86**, 2023–2024

v. Löhneysen H. (2011). Electron-electron interactions and metal-insulator transition in heavily-doped silicon. *Ann. Phys. (Berlin)* **523**, 8-9, 599–611

Lowell J., Munñuz J. S., and Sousa J. (1967). Thermally-induced voltages in the mixed state of type II superconductors. *Phys. Lett.* **24A**, 376–377

Ludoph B. and van Ruitenbeek J. M. (1999). Thermopower of atomic-size metallic contacts. *Phys. Rev. B* **59**, 12290–12293

Lunde A. M. and Flensberg K.(2005). On the Mott formula for the thermopower of non-interacting electrons in quantum point contacts. *J.Phys: Condens. Matter* **17**, 3879–3884

Macdonald D. K. C. (1962). *Thermoelectricity, An Introduction to the Principles*. John Wiley & Sons, New York and London

Macdonald D. K. C., Pearson W. B., and Templeton I. M. (1960). Thermo-electricity at low temperatures VIII. Thermo-electricity of the alkali metals below 2K, *Proc. Royal Soc. London A*, **256**, 334–358

Macdonald D. K. C., Pearson W. B., and Templeton I. M. (1961). Thermo-electricity of lithium alloys at very low temperatures, *Phil. Mag.* **6**, 1431–1437

Macdonald D. K. C., Pearson W. B., and Templeton I. M. (1962). Thermo-electricity at Low Temperatures. IX. The transition metals as solute and solvent. *Proc. R. Soc. A.* **266**, 161–184

Machida Y., Izawa K., Aoki D., Knebel G., Pourret A., and Flouquet J. (2013). Magnetic field driven electronic singularities through metamagnetic phenomena: Case of the heavy fermion antiferromagnet Ce(Ru$_{0.92}$Rh$_{0.08}$)$_2$Si$_2$. *J. Phys. Soc. Jpn.* **82**, 054704

Mackenzie A. P., Julian S. R., Sinclair D. C., and Lin C. T. (1996). Normal-state magnetotransport in superconducting Tl$_2$Ba$_2$CuO$_{6-\delta}$ to millikelvin temperatures. *Phys. Rev. B* **53**, 5848–5855

Mahan G. D. (1998). Good thermoelectrics. *Solid-State Phys.* **41**, 80–157

Mahan G. D., Sales B., and Sharp J. (1997). Thermoelectric materials: New approaches to an old problem. *Phys. Today* **50**(3), 42–47

Mahan G. D. and Sofo J. O. (1996). The best thermoelectric. *Proc. Nat. Acad. Sci., USA* **93**, 7436–7439

Maignan A., Hbert S., Hervieu M., Michel C., Pelloquin D., and Khomskii D. (2003). Magnetoresistance and magnetothermopower properties of Bi/Ca/Co/O and Bi(Pb)/Ca/Co/O misfit layer cobaltites. *J. Phys: Condens. Matter* **15**, 2711–2723

Maki K. (1969). Thermopower in dilute magnetic alloys. *Prog. Theoret. Phys.* **51**, 586–589

Malone L. *et al.* (2012). Thermoelectricity of the ferromagnetic superconductor UCoGe. *Phys. Rev. B* **85**, 024526

Mangez J. H., Issi J.-P., and Heremans J. (1976). Transport properties of bismuth in quantizing magnetic fields. *Phys. Rev. B* **14**, 4381–4385

Marinescu D. C. and Overhauser A. W. (1997). Thermoelectric flux in superconducting rings, *Phys. Rev. B* **55**, 11637–11645

Martin D. L. (1973). Specific heat of copper, silver, and gold below 30 K. *Phys. Rev. B* **8**, 5357–5360

Matsushita Y., Bluhm H., Geballe T. H., and Fisher I. R. (2005). Evidence for charge Kondo effect in superconducting Tl-doped PbTe. *Phys. Rev. Lett.* **94**, 157002

Michaeli K. and Finkelstein A. M. (2009). Fluctuations of the superconducting order parameter as an origin of the Nernst effect. *Europhys. Lett.* **86**, 27007

Mills J. J., Morant R. A., and Wright D. A. (1965). Thermomagnetic effects in pyrolytic graphite. *Brit. J. Appl. Phys.* **16**, 479–481

Mishra S. K., Satpathy S., and Jepsenz O. (1997). Electronic structure and thermoelectric properties of bismuth telluride and bismuth selenide. *J. Phys: Condens. Matter* **9**, 461–470

Miyake K. and Kohno H. (2005). Theory of quasi-universal ratio of Seebeck coefficient to specific heat in zero-temperature limit in correlated metals. *J. Phys. Soc. Jpn.* **74**, 254–258

Molenkamp L. W., Gravier Th., van Houten H., Buijk O. J. A., Mabesoone M. A. A., and Foxon C. T. (1992). Peltier coefficient and thermal conductance of a quantum point contact. *Phys. Rev. Lett.* **68**, 3765–3768

Molenkamp L. W., van Houten H., Beenakker C. W. J., Eppenga R., and Foxon C. T. (1990). Quantum oscillations of the transverse voltage of a channel in the nonlinear transport regime. *Phys. Rev. Lett.* **65**, 1052–1055

Molenkamp, L. W., Staring A. A.M., Alphenaar B.W., van Houten H., and Beenakker C.W. J. (1994). Sawtooth-like thermopower oscillations of a quantum dot in the Coulomb blockade regime. *Semicond. Sci. Technol.* **9**, 903–906

Mooser E. and Woods S. B. (1955). Thermoelectric power of germanium at low temperatures; *Phys. Rev.* **97**, 1721

Morelli D. T., Jovovic V,. and Heremans J. P. (2008). Intrinsically minimal thermal conductivity in cubic I-V-VI2 semiconductors. *Phys. Rev. Lett.* **101**, 035901

Mori T. and Inokuchi H. (1988). Thermoelectric power of organic superconductors calculation on the basis of the tight-binding theory. *J. Phys. Soc. Jpn.* **57**, 3674–3677

Mortensen K., Conwell E. M., and Fabre J. M. (1983). Thermopower studies of a series of salts of tetramethyltetrathiafulvalene [(TMTTF)$_2$X, X=Br, ClO$_4$, NO$_3$, SCN, BF$_4$, AsF$_6$, and PF$_6$]. *Phys. Rev. B* **28**, 5856–5862

Mortensen K. and Engler E. M. (1984). Conductivity and thermopower studies of bis-tetramethyltetraselenafulvalenium hexafluorophosphide, bis-tetramethyltetrathiafulvalenium hexafluorophosphide, and their solid solutions, (TMTSF$_{1-x}$TMTTF$_x$)2PF$_6$ . *Phys. Rev. B* **29**, 842–850

Mott N. F. and Jones H. (1936). *The Theory of the Properties of Metals and Alloys*. Clarendon Press, Oxford

Müller K. A. and Burkard H. (1979). SrTiO$_3$: An intrinsic quantum paraelectric below 4 K. *Phys. Rev. B* **19**, 3593–3602

Mydosh J. A. and Oppeneer P. M. (2011). Colloquium: hidden order, superconductivity, and magnetism: The unsolved case of URu$_2$Si$_2$. *Rev. Mod. Phys.* **83**, 1301

Nakamae S., Behnia K., Mangkorntong N., Nohara M., Takagi H., Yates S. J. C., and Hussey N. E. (2003). Electronic ground state of heavily overdoped nonsuperconducting La$_{2-x}$Sr$_x$CuO$_4$. *Phys. Rev. B* **68**, 100502(R)

Nakamura Y. and Uchida S. (1993). Anisotropic transport properties of single-crystal La$_{2-x}$Sr$_x$CuO$_4$: Evidence for the dimensional crossover. *Phys. Rev. B* **47**, 8369–8373

Nam M.-S., Ardavan A., Blundell S. J., and Schlueter J. A. (2007). Fluctuating superconductivity in organic molecular metals close to the Mott transition. *Nature* **449**, 584–587

Narduzzo A. *et al.* (2008). Violation of the isotropic mean free path approximation for overdoped La$_{2-x}$Sr$_x$CuO$_4$. *Phys. Rev. B* **77**, 220502(R)

Newman P. F. and Holcomb D. F. (1983). Metal-insulator transition in Si:As. *Phys. Rev. B* **28**, 638–640

Nielsen P. E. and Taylor P. L. (1970). New effect in the electron-diffusion thermopower of metals. *Phys. Rev. Lett.* **25**, 371–373

Nielsen P. E. and Taylor P. L. (1974). Theory of thermoelectric effects in metals and alloys. *Phys. Rev. B* **10**, 4061–4070

Nolas G. S., Sharp J., and Goldsmid J. (2001). *Thermoelectrics: Basic Principles and New Materials Developments*. Springer-Verlag, Berlin

Nordheim, L., and Gorter C. J. (1935). Bemerkungen über thermokraft und widerstand. *Physica*, **2**, 383–390

Oberli L., Manuel A. A., Sachot R., Descouts P., and M. Peter M. (1985). Fermi surface of lithium studied by positron annihilation, *Phys. Rev. B* **31**, 6104–6107

Obertelli S. D., Cooper J. R., and Tallon J. L. (1992). Systematics in the thermoelectric power of high-T$_c$ oxides *Phys. Rev. B* **46**, 14928–14931

Oganesyan V. and Ussishkin I. (2004). Nernst effect, quasi-particles, and d-density waves in cuprates. *Phys. Rev. B* **70**, 054503

Ohta S., Nomura T., Ohta H., Hirano M., Hosono H., and Koumoto K. (2005). Large thermo-electric performance of heavily Nb-doped SrTiO$_3$ epitaxial film at high temperature. *Appl. Phys. Lett.* **87**, 092108

Oji j. (1984). Thermomagnetic effects in two-dimensional electron systems. *J. Phys. C* **17**, 3059–3066

Ōkuda T., Nakanishi K., Miyasaka S., and Tokura Y. (2001). Large thermoelectric response of metallic perovskites: $Sr_{1-x}La_xTiO_3$. *Phys. Rev. B* **63**, 113104

Ong N. P. (1991). Geometric interpretation of the weak-field Hall conductivity in two-dimensional metals with arbitrary Fermi surface. *Phys. Rev. B* **43**, 193–201

Onsager L. (1931a). Reciprocal relations in irreversible processes. I. *Phys. Rev.* **37**, 405–426

Onsager L. (1931b). Reciprocal relations in irreversible processes. II. *Phys. Rev.* **38**, 2265–2279

Paglione J. *et al.* (2003). Field-induced quantum critical point in $CeCoIn_5$. *Phys. Rev. Lett.* **91**, 246405

Palstra T. T. M., Batlogg B., Schneemeyer L. F., and Waszczak J. V. (1990). Transport entropy of vortex motion in $YBa_2Cu_3O_7$. *Phys. Rev. Lett.* **64**, 3090–3093

Palstra T. T. M. *et al.* (1997). Transport mechanisms in doped $LaMnO_3$: Evidence for polaron formation. *Phys. Rev. B* **56**, 5104–5107

Parker D., Chen X., and Singh D. J. (2013). High three-dimensional thermoelectric performance from low-dimensional bands. *Phys. Rev. Lett.* **110**, 146601

Parkinson D. H. and Quarrington J. E. (1954). The molar heats of lead sulphide, selenide and telluride in the temperature range 20 K to 260 K. *Proc. Phys. Soc. A* **67**, 569–579

Peets D. C. *et al.* (2007). $Tl_2Ba_2CuO_{6+\delta}$ brings spectroscopic probes deep into the overdoped regime of the high-$T_c$ cuprates. *New J. Phys.* **9**, 28

Pei Y., LaLonde A., Iwanaga S., and Snyder G. J. (2011a). High thermoelectric figure of merit in heavy hole dominated PbTe. *Energy Environ. Sci.* **4**, 2085–2089

Pei Y., Shi X., LaLonde A., Wang H., Chen L., and Snyder G. J. (2011b). Convergence of electronic bands for high performance bulk thermoelectrics. *Nature* **473**, 66–69

Pfau H., Daou R., Brando M., and Steglich F. (2012). Thermoelectric transport across the metamagnetic transition of $CeRu_2Si_2$ *Phys. Rev. B* **85**, 035127

Pourret A., Aubin H., Lesueur J., Marrache-Kikuchi C. A., Bergé L., Dumoulin L., and Behnia K. (2006). Observation of the Nernst signal generated by fluctuating Cooper pairs. *Nat. Phys.* **2**, 683–686

Pourret A., Aubin H., Lesueur J., Marrache-Kikuchi C. A., Bergée L., Dumoulin L., and Behnia K. (2007). Length scale for the superconducting Nernst signal above $T_c$ in $Nb_{0.15}Si_{0.85}$. *Phys. Rev. B* **76**, 214504

Pourret A., Spathis P., Aubin H., and Behnia K. (2009). Nernst effect as a probe of superconducting fluctuations in disordered thin films, *New J. Phys.* **11**, 055071

Price P. J. (1956). Theory of transport effects in semiconductors: thermoelectricity. *Phys. Rev.* **104**, 1223–1239

Proust C., Boaknin E., Hill R. W., Taillefer L., and Mackenzie A. P.(2002). Heat transport in a strongly overdoped cuprate: Fermi liquid and a pure d-wave BCS superconductor. *Phys. Rev. Lett.* **89**, 147003

Randles D. L. and Springford M. (1976). De Haas-van Alphen effect in alkali metal dispersions and the Fermi surface of lithium, *J. Phys. F: Met. Phys.* **6**, 1827–1844

Reddy P., Jang S.-Y., Segalman R. A., and Majumdar A. (2007). Thermoelectricity in molecular junctions. *Science* **315**, 1568–1571

Ri H.-C., Gross R., Gollnik F., Beck A., and Huebener R. P. (1994). Nernst, Seebeck, and Hall effects in the mixed state of $YBa_2Cu_3O_{7-\delta}$ and $Bi_2Sr_2CaCu_2O_8$ thin films: A comparative study. *Phys. Rev. B* **50**, 3312–3330

Roberts R. B. (1977). The absolute scale of thermoelectricity. *Phil. Mag.* **6**, 1431–1437

Robinson J. E. (1967). Thermoelectric power in the nearly-free-electron model. *Phys. Rev.* **161**, 531–539

Robinson J. E. and Dow J. D. (1968). Electron-phonon interactions in solid alkali metals. I. scattering and transport coefficients. *Phys. Rev.* **171**, 815–826

Rosenbaum T. F., Andres K., Thomas G. A., and Bhatt R. N. (1980). Sharp metal-insulator transition in a random solid *Phys. Rev. Lett.* **45**, 1723–1726

Rourke P. M. C. *et al.* (2010). A detailed de Haas-van Alphen effect study of the overdoped cuprate $Tl_2Ba_2CuO_{6+d}$, *New J. Phys.* **12**, 105009

Rubio G., Agraït N., and S. Vieira S. (1996). Atomic-sized metallic contacts: mechanical properties and electronic transport. *Phys. Rev. Lett.* **76**, 2302–2305

Rumbo E. R. (1976). Transport properties of very pure copper and silver below 8.5 K. *J. Phys. F: Met. Phys.* **6**, 85–98

Sakai A., Kanno T., Yotsuhashi S., Adachi H., and Tokura Y. (2009). Thermoelectric properties of electron-doped $KTaO_3$. *Jap. J. Appl. Phys.* **48**, 097002

Sakurai J., Hasegawa K., Menovsky A. A., and Schweizer J. (1996). Thermoelectric power on single crystals of $URu_2Si_2$. *Sol. State Commun.* **97**, 689–691

Sakurai J. and Isikawa Y. (2005). Correlation of initial slope of thermoelectric power and electronic specific-heat constant for solid solutions based on Ce and Eu compounds. *J. Phys. Soc. Jpn.* **74**, 1926–1929

Salamon M. B. and Jaime M. (2001). The physics of manganites: Structure and transport. *Rev. Mod. Phys.* **73**, 583–628

Sarrao J. L. and Thompson J. D. (2007). Superconductivity in cerium- and plutonium-based '115' materials. *J. Phys. Soc. Jpn.* **76**, 051013

Scheidemantel T. J., Ambrosch-Draxl C., Thonhauser T., Badding J. V., and Sofo J. O. (2003). Transport coefficients from first-principles calculations. *Phys. Rev. B* **68**, 125210

Scholz K., Jandl P., Birkholz J., and Dashevskii J. M. (1994). Infinite stage Ettingshausen cooling in Bi-Sb alloys. *J. Appl. Phys.* **75**, 5406–5408

Schooley J. F., Hosler W. R., and Cohen M. L. (1964). Superconductivity in Semiconducting $SrTiO_3$. *Phys. Rev. Lett.* **12**, 474–475

Schweitzer H. and Czycholl G. (1991). Resistivity and thermopower of heavy-fermion systems. *Phys. Rev. Lett.* **67**, 3724–3727

Sehlin S. R., Anderson H. U., and Sparlin D. M. (1995). Semi-empirical model for the electrical properties of $La_{1-x}Ca_xCoO_3$. *Phys. Rev. B* **52**, 11681–11689

Seol J. H., Moore A. L., Saha S. K., Zhou F., Shi L., Ye Q. L., Scheffler R., Mingo N., and T. Yamada T. (2007). Measurement and analysis of thermopower and electrical conductivity of an indium antimonide nanowire from a vapor-liquid-solid method. *J. Appl. Phys.* **101**, 023706

Serbyn M. N., Skvortsov M. A., Varlamov A. A., and Galitski V. (2009). Giant Nernst effect due to fluctuating cooper pairs in superconductors. *Phys. Rev. Lett.* **102**, 067001

Shikano M. and Funahashi R. (2003). Electrical and thermal properties of single-crystalline $(Ca_2CoO_3)_{0.7}CoO_2$ with a $Ca_3Co_4O_9$ structure. *Appl. Phys. Lett.* **82**, 1851–1953

Shoenberg D. (1984). *Magnetic Oscillations in Metals*. Cambridge University Press, Cambridge

Silvera I. S. (1997). Bose-Einstein condensation. *Am. J. Phys.* **65**, 570–574

Singh D. J. (2000). Electronic structure of $NaCo_2O_4$. *Phys. Rev. B* **61**, 13397–13402

Singh D. J. (2010). Doping-dependent thermopower of PbTe from Boltzmann transport calculations. *Phys. Rev. B* **81**, 195217

Sitter H., Lischka K., and Heinrich H. (1977). Structure of the second valence band in PbTe. *Phys. Rev. B* **16**, 680–687

Sivan U. and Imry Y. (1986). Multichannel Landauer formula for thermoelectric transport with application to thermopower near the mobility edge. *Phys. Rev. B* **33**, 551–558

Slachter A., Bakker F. L., Adam J. P., and van Wees B. J. (2010). Thermally driven spin injection from a ferromagnet into a non-magnetic metal. *Nat. Phys.* **6**, 879–883

Small J. P., Perez K. M., and Kim P. (2003). Modulation of thermoelectric power of individual carbon nanotubes. *Phys. Rev. Lett.* **91**, 256801

Smith A. C., Janak J. F., and Adler R. B. (1967). *Electron Conduction in Solids*. McGraw-Hill, New York

Smith G. E. and Wolfe R. (1962). Thermoelectric properties of bismuth-antimony alloys. *J. Appl. Phys.* **33**, 841–846

Snyder G. J. and Tobere E. S. (2008). Complex thermoelectric materials. *Nat. Mater.* **7**, 105–114

Solomon P. R. and Otter Jr. F. A. (1967). Thermomagnetic Effects in Superconductors. *Phys. Rev.* **164**, 608–618

Sommerfeld A. and Frank N. H. (1931) The statistical theory of thermoelectric, galvano- and thermomagnetic phenomena in metals. *Rev. Mod. Phys.* **3**, 1–42

Sondheimer E. H. (1956a). The Kelvin relations in thermo-electricity. *Proc. Roy. Soc.* London **A234**, 391–398

Sondheimer E. H. (1956b). Electron-phonon equilibrium and the transport phenomena in metals at low temperatures, *Can. J. Phys.*, **34**, 1246–1255

Sonntag J. (2010). The effect of band edges on the Seebeck coefficient, *J. Phys. Condens. Matter* **22**, 235501

Steglich F., Aarts J., Bredl C. D., Lieke W., Meschede D., Franz W., and Schäfer H. (1979). Superconductivity in the presence of strong pauli paramagnetism: CeCu2Si2, *Phys. Rev. Lett.* **43**, 1892–1896

Stephen M. J. (1966). Galvanomagnetic and related effects in type-II superconductors. *Phys. Rev. Lett.* **16**, 801–803

Stewart G. R. (2011). Superconductivity in iron compounds. *Rev. Mod. Phys.* **83**, 1589

Streda P. (1989). Quantised thermopower of a channel in the ballistic regime. *J. Phys. Condens. Matter* **1**, 1025–1027

Sugihara K. (1969). Thermomagnetic effects in bismuth. II. Nernst-Ettingshausen Effect. *J. Phys. Soc. Jpn.* **27**, 362–370

Szilard L. (1929). Uber die entropieverminderung in einem thermodynamischen system be eingriffen intelligenter wesen, *Zeitschrift für Physik* **53**, 840–856

Tafti F. F., Laliberté F., Dion M., Gaudet J., Fournier P., and Taillefer L. (2014). Nernst effect in the electron-doped cuprate superconductor PCCO: Superconducting fluctuations, upper critical field $H_{c2}$, and the origin of the $T_c$ dome. *Phys. Rev. B* **90**, 024519

Takada K., Sakurai H., Takayama-Muromachi E., Izumi F., Dilanian R. A., and Sasaki T. (2003). Superconductivity in two-dimensional CoO2 layers. *Nature* **422**, 53–55

Terasaki I. (2011). High-temperature oxide thermoelectrics. *J. App. Phys.* **110**, 053705

Terasaki I., Sasago Y., and Uchinokura K. (1997). Large thermoelectric power in $NaCo_2O_4$ single crystals. *Phys. Rev. B* **56**, R12685–R12687

Terasaki I., Tsukada I., and Iguchi Y. (2002). Impurity-induced transition and impurity-enhanced thermopower in the thermoelectric oxide $NaCo_{2-x}Cu_xO_4$. *Phys. Rev. B* **65**, 195106

Tieke B., Fletcher R., Maan J. C., Dobrowolski W., Mycielski A., and Wittlin A. (1996). Magnetothermoelectric properties of the degenerate semiconductor HgSe:Fe *Phys. Rev. B* **54**, 10565–10574

Tieke B., Fletcher R., Zeitler U., Geim A. K., Henini M., and Maan J. C. (1997). Fundamental relation between electrical and thermoelectric transport coefficients in the quantum Hall regime. *Phys. Rev. Lett.* **78**, 4621–4624

Tieke B., Fletcher R., Zeitler U., Henini M., and Maan J. C. (1999). Thermopower measurements of the coupling of phonons to electrons and composite fermions. *Phys. Rev. B* **58**, 2017–2025

Tokura Y. *et al.* (1993). Filling dependence of electronic properties on the verge of metal Mott-insulator transition in $Sr_{1-x}La_xTiO_3$. *Phys. Rev. Lett.* **70**, 2126–2129

Trego A. L. and Mackintosh A. R. (1968). Antiferromagnetism in chromium alloys. II. Transport properties *Phys. Rev.* **166**, 495–506

Tsuei C. C. and Kirtley J. R. (2000). Pairing symmetry in cuprate superconductors, *Rev. Mod. Phys.* **72**, 969–1016

Uchida K. *et al.* (2010). Spin Seebeck insulator. *Nature Mater.* **9** 894–897

Uchida K. *et al.* (2011). Thermoelectric response in the incoherent transport region near Mott transition: The case study of $La_{1-x}Sr_xVO_3$. *Phys. Rev. B* **83**, 165127

Uchida M. *et al.* (2011). Thermoelectric response in the incoherent transport region near Mott transition: The case study of $La_{1-x}Sr_xVO_3$. *Phys. Rev. B* **83**, 165127

Uchida K., Takahashi S., Harii K., Ieda J., Koshibae W., Ando K., Maekawa S., and Saitoh E. (2008). Observation of the spin Seebeck effect. *Nature* **455**, 778–781

Uher C. and Pratt W. P. Jr. (1978). Thermopower measurements on bismuth from 9K down to 40 mK. *J. Phys. F* **8**, 1979–1989

Urayama H. *et al.* (1988). Valence state of copper atoms and transport property of an organic superconductor, $(BEDT-TTF)_2Cu(NCS)_2$, measured by ESCA, ESR, and thermoelectric power. *Chem. Lett.* **17**, 1057–1060

Ussishkin I., Sondhi S. L., and Huse D. A. (2002). Gaussian superconducting fluctuations, thermal transport, and the Nernst effect. *Phys. Rev. Lett.* **89**, 287001

Valenzuela S. O. and Tinkham M. (2006). Direct electronic measurement of the spin Hall effect. *Nature* **442**, 176–179

Vidal F. (1973). Low-frequency ac measurements of the entropy flux associated with the moving vortex lines in a low-$\kappa$ type-II superconductor. *Phys. Rev. B* **8**, 1982–1993

Vignolle B. *et al.*(2008). Quantum oscillations in an overdoped high-Tc superconductor. *Nature* **455**, 952–955

Wang Y., Li L., and Ong N. P. (2006). Nernst effect in high-$T_c$ superconductors. *Phys. Rev. B*, **73**, 024510

Wang Y., Rogado N. S., Cava R. J., and Ong N. P. (2003). Spin entropy as the likely source of enhanced thermopower in $Na_xCo_2O_4$. *Nature* **423**, 425

Wang Z., Wang S., Obukhov S., Vast N., Sjakste J., Tyuterev V., and Mingo N. (2011). Thermoelectric transport properties of silicon: Toward an ab initio approach. *Phys. Rev. B* **83**, 205208

Weber L. and Gmelin E. (1991). Transport properties of silicon. *Appl. Phys. A* **53**, 136–140

van Wees B. J., van Houten H., Beenakker C. W. J., Williamson J. G., Kouwenhoven L. P., van der Marel D., and Foxon C. T. (1988). Quantized conductance of point contacts in a two-dimensional electron gas. *Phys. Rev. Lett.* **60**, 848–850

Wei P., Bao W., Pu Y., Lau C. N., and Shi J. (2009). Anomalous thermoelectric transport of dirac particles in graphene. *Phys. Rev. Lett.* **102**, 166808

Wendling N., Chaussy J., and Mazuer J. (1993). Thin gold wires as reference for thermoelectric power measurements of small samples from 1.3 K to 350 K, *J. Appl. Phys.* **73**, 2878–2882

Wiese J. R. and Muldawer L. (1960). Lattice constants of $Bi_2Te_3$–$Bi_2Se_3$ solid solution alloys *J. Phys. Chem. Solids* **15**, 13–16

Wolfe R. and Smith G. E. (1962). Effects of a magnetic field on the thermoelectric properties of a bismuth-antimony alloy, *Appl. Phys. Lett.* **1**, 5–7

Wolff P. A. (1964). Matrix elements and selection rules for the two-band model of bismuth. *J. Phys. Chem. Solids* **25**, 1057–1068

Woollam J. A. (1971). Graphite carrier locations and quantum transport to 10 T (100 kG). *Phys. Rev. B* **3**, 1148–1159

Wu S. and Lebed A. G. (2010). Unification theory of angular magnetoresistance oscillations in quasi-one-dimensional conductors. *Phys. Rev. B* **82**, 075123

Wu W., Lee I. J., and Chaikin P. M. (2003). Giant Nernst effect and lock-in Currents at magic angles in $(TMTSF)_2PF_6$. *Phys. Rev. Lett.* **91**, 056601

Wu J. and Leighton C. (2003). Glassy ferromagnetism and magnetic phase separation in $La_{1-x}Sr_xCoO_3$. *Phys. Rev. B* **67**, 174408

Wu W., Ong N. P., and Chaikin P. M. (2005). Giant angular-dependent Nernst effect in the quasi-one-dimensional organic conductor $(TMTSF)_2PF_6$. *Phys. Rev. B* **72**, 235116

Xu Z. A., Ong N. P., Wang Y., Kakeshita T., and Uchida S. (2000). Vortex-like excitations and the onset of superconducting phase fluctuations in underdoped $La_{2-x}Sr_xCuO_4$. *Nature* **406**, 486–488

Yamamoto A., Hu W. -Z., and Tajima S. (2000). Thermoelectric power and resistivity of $HgBa_2CuO_{4+\delta}$ over a wide doping range. *Phys. Rev. B* **63**, 024504

Yan Z. (2000). General thermal wavelength and its applications. *Eur. J. Phys.* **21**, 625–631

Yim W. M. and Amith A. (1972). Bi-Sb alloys for magneto-thermoelectric and thermomagnetic cooling. *Solid-State Elect.* **15**, 1141–1165

Yu R. C. *et al.* (1988). Observations on the thermopower of the high-$T_c$ superconductors. *Phys. Rev. B* **37**, 7963–7966

Yu R. *et al.* (1991). Anisotropic thermopower of the organic superconductor $\kappa$-(BEDT-TTF)$_2$Cu[N(CN)$_2$]Br. *Phys. Rev. B* **44**, 6932–6936

Zavaritsky N. V. (1987). Phonon-drag effect in three- and two-dimensional electron systems. *Pramāna– J. Phys.* **28**, 489–501

Zener C. (1951). Interaction Between the d shells in the transition metals. *Phys. Rev.* **81**, 440–444

Zhang H., Liu C. -X., Qi X. -L., Dai X., Fang Z., and Zhang S.-C. (2009). Topological insulators in $Bi_2Se_3$, $Bi_2Te_3$ and $Sb_2Te_3$ with a single Dirac cone on the surface. *Nat. Phys.* **5**, 438–442

Zhu Z., Fauqué B., Fuseya Y., and Behnia K. (2011). Angle-resolved Landau spectrum of electrons and holes in bismuth. *Phys. Rev. B* **84**, 115137

Zhu Z., Fauqué B., Malone L., Antunes A. B., Fuseya Y., and Behnia K. (2012). Landau spectrum and twin boundaries of bismuth in the extreme quantum limit. *Proc. Natl. Acad. Sci., USA* **109**, 14813–14818

Zhu Z., Hassinger E., Xu Z., Aoki D., Flouquet J., and Behnia K. (2009). Anisotropic inelastic scattering and its interplay with superconductivity in $URu_2Si_2$. *Phys. Rev. B* **80**, 172501

Zhu Z., Yang H., Fauqué B., Kopelevich Y., and Behnia K. (2010). Nernst effect and dimensionality in the quantum limit. *Nat. Phys.* **6**, 26–29

Ziman J. M. (1960). *Electrons and Phonons*. Oxford University Press, Oxford

Ziman J. M. (1961). The ordinary transport properties of the noble metals. *Adv. Phys.* **10**, 1–56

Ziman J. M. (1964). *Principles of the Theory of solids*. Cambridge University Press, Cambridge

Zlatić V., Horvatić B., Milat I., Coqblin B., Czycholl G., and Grenzebach C. (2003). Thermoelectric power of cerium and ytterbium intermetallics. *Phys. Rev. B* **68**, 104432

Zlatić V., Monnier R., Freericks J. K., and Becker K. W. (2007). Relationship between the thermopower and entropy of strongly correlated electron systems. *Phys. Rev. B* **76**, 085122

Zuev Y. M., Chang W., and Kim P. (2009). Thermoelectric and magnetothermoelectric transport measurements of graphene. *Phys. Rev. Lett.*, **102**, 096807

# Index